Microbiota

Current Research and Emerging Trends

https://doi.org/10.21775/9781910190937

Edited by

Takashi Matsumoto[1]

and

Yoshio Yamaoka[1,2]

[1]Department of Environmental and Preventive Medicine
Faculty of Medicine, Oita University
Yufu
Japan

[2]Department of Medicine
Gastroenterology and Hepatology Section
Baylor College of Medicine
Houston, TX
USA

Caister Academic Press

Caister Academic Press
Norfolk, UK

www.caister.com

British Library Cataloguing-in-Publication Data
A catalogue record for this book is available from the British Library

ISBN: 978-1-910190-93-7 (paperback)
ISBN: 978-1-910190-94-4 (ebook)

Contents

Current Books of Interest

Preface

Since the existence of trillions of microbes in our body has been described, human microbiota research has become a recent trend in the field of medical science. In particular, the development of high-throughput sequencing techniques and bioinformatics has made a great contribution to understanding the dynamic relationship between these microbes and human health, such as in inflammatory bowel disease, type 2 diabetes and obesity. In fact, over 3500 articles were published in 2017 and deposited on PubMed. These studies reveal that microbiota may contribute to many different types of diseases, early development and ageing. Here, we provide a summary of recent human microbiota studies and describe the current understanding of healthy microbial diversity, its functional roles in development and the potential of probiotic therapy in future studies of human microbiota.

'Microbiota' refers to the microorganisms in a particular environment and also the microbial community inhabiting our bodies. The 'microbiome' is an ensemble of the genomic material of these microorganisms. This genomic information can be utilized to understand the existence of a microbial community, using current analytical approaches. The recent development of next-generation sequencing methods has greatly facilitated the study of the microbiota. In Chapter 1, Nuria Salazar *et al.* describe non-culture methods and high-throughput sequencing techniques for microbiome study and provide an overview of the association of microbiota with dietary patterns and lifestyle factors and evidence of its influence on health status in the elderly.

Alteration of the microbiota during early life is regarded as one of the factors that increases the risk of non-communicable diseases and metabolic disorders. In Chapters 2 and 3, Yuichiro Yamashiro *et al.* provide a comprehensive understanding of the aberrant bacterial community 'dysbiosis' in early life and its contribution to the development of the immune system and metabolic conditions and its influence on the developmental origins of health and disease. Furthermore, they describe the efficacy of probiotics in treating paediatric illnesses such as preterm low-birthweight, allergy associated with caesarean section delivery, mucositis and child obesity. They briefly summarise the beneficial effects of probiotic supplementation in promoting a more favorable microbiota composition and the progress of current probiotic applications.

There is a complex crosstalk between microbiota and the host cells. Metaproteomics is a novel omics approach that can be used for the large-scale characterization of the entire protein complement of the environmental microbiota. Both the human and the microbiota proteins can provide information on the functional aspects of the proteolytic events in the gut environment. Together with evidence from the complementary approaches of metagenomics and metatranscriptomics, metaproteomics may provide new insights into the role of the microbiome. General knowledge of these processes and approaches are described in Chapter 4 by Bernardo Petriz *et al.*

Recently, it has been identified that the gut microbiota contributes to biochemical signalling from the gastrointestinal tract to the central nervous system as bidirectional communication between the central and the enteric nervous systems. In Chapter 5, Surajit Pathak *et al.* describe how this dynamic relationship plays an important role in normal brain development and dysfunction, leading to several diseases. Gaining a better understanding of this bidirectional pathway can provide an insight into innovative and therapeutic strategies for neurodevelopmental and behavioral disorders. Not only microbiota in the gut, but also microbiota in the oral cavity, have a key role to play in maintaining our health status. A diverse population of microorganisms exists in our oral cavity. In Chapter 6, Surajit Pathak *et al.* describe how recent studies have suggested that oral pathogens are involved in the production of chronic inflammatory responses and that these may contribute to the development of neurodegenerative diseases such as Parkinson's disease and Alzheimer's disease. There is a need to expand our knowledge of the composition and function of the oral microbiota, which contribute to our brain health.

Changes in the structure of the gut microbiota contribute to our health and to diseases. However, it is still unclear whether or not the ecological niche of a microbial community is associated with the onset of primary tumours in other organs. In Chapter 7, Vo Phuoc Tuan *et al.* describe the structure of the microbiota in the gastrointestinal tract, particularly in the esophagus, stomach and colon, providing a comprehensive summary of current studies in this area and how the composition of the microbiota changes as a result of diseases such as gastric cancer. In addition, they describe the potential application of therapeutic approaches targeting microbiota and discuss future trends in microbiota studies.

We hope that this book will help you to understand the current research and emerging trends in the study of microbiota.

Takashi Matsumoto and Yoshio Yamaoka,
Oita, Japan

Microbiome: Effects of Ageing and Diet

1

Nuria Salazar[1,2], Sonia González[2,3], Alicja M. Nogacka[1,2], David Rios-Covián[1,2], Silvia Arboleya[1,2], Miguel Gueimonde[1,2] and Clara G. de los Reyes-Gavilán[1,2*]

[1]Department of Microbiology and Biochemistry of Dairy Products, Instituto de Productos Lácteos de Asturias, Spanish Research Council (IPLA-CSIC), Villaviciosa, Spain.
[2]Diet, Microbiota and Health Research Team, Instituto de Investigación Sanitaria del Principado de Asturias (ISPA), Oviedo, Spain.
[3]Department of Functional Biology, Faculty of Medicine, University of Oviedo, Oviedo, Spain.

*Correspondence: greyes_gavilan@ipla.csic.es

https://doi.org/10.21775/9781910190937.01

Abstract
The microbial community inhabiting our intestine, known as 'microbiota', and the ensemble of their genomes (microbiome) regulate important functions of the host, being essential for health maintenance. The recent development of next-generation sequencing (NGS) methods has greatly facilitated the study of the microbiota and has contributed to evidence of the strong influence exerted by age and diet. However, the precise way in which the diet and its components modify the functionality of the intestinal microbiome is far from being completely known. Changes in the intestinal microbiota occur during ageing, frequently accompanied by physiological changes of the digestive tract, modification of dietary patterns and impairment of the immune system. Establishing nutritional strategies aiming to counterbalance the specific alterations taking place in the microbiota during ageing would contribute to improved health status in the elderly. This chapter will analyse changes appearing in the intestinal microbiota from adulthood to old age and their association with dietary patterns and lifestyle factors.

Introduction
The microbiota is the pool of commensal microorganisms colonizing our body and includes bacteria, archaea, viruses and unicellular eukaryotes. The largest microbial population is found in the human gut; in particular, the colon is the most densely populated area, with a total microbial content estimated as about 10^{14} cells (Hooper and Gordon, 2001). The ensemble of genes contained by this microbiota is known as the 'intestinal microbiome' and they encode for multiple functions, some of which strongly influence the host physiology

and metabolism. The gut microbiota and its metabolites interact with the host at different levels and its correct composition and functionality is essential for maintaining a 'healthy status' through life. The advent of culture-independent techniques, particularly the more recent development and cost affordability of DNA next-generation sequencing (NGS) and the omics technologies (metatranscriptomics, metaproteomics and metabolomics), have made possible the study of the microbiota composition and its functional capacity. The latest scientific advances have evidenced that the microbiota is strongly influenced by diet and age, whilst some important aspects of its functionality still remain unknown. Short-chain fatty acids (SCFA; acetate, propionate and butyrate) are the most abundant metabolites produced by the microbiota as end products from the fermentation of indigestible carbohydrates and proteins in the colon. It is known that their production is influenced by dietary patterns. These compounds affect the physiology of the colon, contribute to shape the intestinal microbial environment, serve as energy source to host cells and influence the host metabolism and physiology (Rios-Covian *et al.*, 2016).

The gut microbiota colonization starts early after birth and even a primary fetal colonization may occur. The diversity, stability and complexity of this microbiota evolves until the approximate age of three years, remains without noticeable variations in healthy adults and becomes unstable again in the elderly. Changes in the intestinal microbiota occurring during ageing are concomitant with a general physiological decline, anatomical modifications, cognitive impairment and a deregulation of the immune system. This, together with variations in dietary habits promoted by social changes (solitude, living in long-term residential institutions, etc.), anatomical impairment and lower energy requirements due to changes in the whole body composition, increases the risk of nutritional deficiencies and malnutrition in the elderly. However, there still are a limited number of studies describing the composition of the microbiota in elderly individuals and even less about its functionality.

Lifespan is increasing all over the world, with this trend being more pronounced in industrialized countries. According to a recent report from the United Nations (2015), the population aged over 60 years is the fastest growing, and more than 20% of the world's population will be over the age of 60 years by 2050. The increase in life expectancy is one of the most relevant social and medical achievements of the last two centuries. However, the rise in health-care costs associated with recurrent disease and disability frequently occurring in the elderly represents an economic burden and a sustainability challenge for our current social structure. Therefore, it is urgent that scientists and medical professionals seek an understanding of how to maintain health throughout the ageing process whereas social and economic policies should be implemented in order to translate scientific advances to the general population, with the final perspective of contributing to healthy ageing and ultimately to a sustainable welfare society. Besides physical activity and lifestyle, nutrition is one of the most important factors playing a key role in healthy ageing. The possibility of shaping the microbiota through diet in order to beneficially affect host health is a priority within this multidisciplinary task.

The present chapter aims to analyse from a global perspective the current state of knowledge about the effect of age and nutritional status on the gut microbiota during the ageing process and the nutritional and lifestyle strategies that contribute to favour healthy ageing of the body and its microbiota. This analysis is approached from the perspective of the most recent techniques available for the study of the composition and functionality of the intestinal microbiota, considering the physiological, immune and nutritional states that influence the global health status of the individual.

The physiology and the immune status in the elderly

With the advancement of age there is a normal and continuous decline in physiological functions, which constitutes the so-called 'senescence' process. This natural process, not a disease, is multifactorial and several inter-related body functions are affected, including metabolic alterations, inflammation, macromolecular damage and reduced protein homeostasis, among others (Kennedy *et al.*, 2014). The impairment of these functions renders the individuals more prone to different conditions such as infectious and digestive diseases, bone injuries or mental health problems. The incidence of these chronic diseases represents a serious societal concern related to the ageing of the population. As an example, mental health-related problems, such as dementia, are expected to double in the next three decades (Dolgin, 2016), giving an idea of the extraordinary magnitude of the challenges associated with an aged population and underlining the need for immediate action. It is thus of great importance to differentiate between normal physiological decline and the changes related to pathological processes with high incidence in aged populations. In this context, the term 'senescence' is often used to refer to non-pathological biological and physiological processes strictly dependent on age, whereas 'ageing' would refer to changes, physiological and pathological, associated with the passage of time (Rowe, 1997; Troen, 2003).

Different body functions are affected by the ageing process, among them immune deregulation (Mazari and Lesourd, 1998), and changes in the intestinal microbiota (Odamaki *et al.*, 2016) and gastrointestinal physiology (Remond *et al.*, 2015) are well recognized. The age-associated changes in the immune function are multifaceted and involve both cellular and humoral immunity. These changes include a reduced activity of immune cells, such as natural killer cells and T cells, a defective response to mitogen-induced proliferative responses, or an increased level of circulating pro-inflammatory cytokines (Jing *et al.*, 2007; Candore *et al.*, 2008, 2010; Vallejo, 2011; Salazar *et al.*, 2013). These alterations, often referred to as 'immunosenescence' (Larbi *et al.*, 2008; Monti *et al.*, 2017), may contribute to explain the increased disease risk observed in elderly people. Moreover, these changes in immune functions may promote neurological disorders. As an example, the existence of a pro-inflammatory status in senior population, which includes also neuroinflammation (Layé *et al.*, 2011), may lead to cognitive decline and neurodegenerative diseases (Wyss-Coray, 2016). This potential primary role of inflammation in different age-related diseases has led to the proposal of the 'inflammaging' theory (Franceschi *et al.*, 2000).

Concomitantly with the above-mentioned immune changes, senescence is also related to difficulties in maintaining physiological and metabolic homeostasis. At the gastrointestinal tract level, changes from mouth to colon occur and go from reduced production of saliva (xerostomia) and other digestive secretions to impaired absorption of nutrients and slower gastrointestinal transit (Culross, 2008; Gonsalves *et al.*, 2008; Rémond *et al.*, 2015). These gastrointestinal changes, together with the loss of taste, induce changes in food preferences and dietary patterns, increasing the risk of nutritional deficiencies and malnutrition (Salazar *et al.*, 2017). As a consequence, there is a decrease of lean and bone mass, a phenomenon known as 'sarcopenia' (Beaufrère and Morio, 2000), which further compromises the homeostasis of the individual.

To summarize, the ageing process is linked to several changes in dietary patterns, immune senescence, impaired gastrointestinal physiology and a concomitant decline in cognitive function. This process is also related to changes in microbiota functionality and composition (dysbiosis), all together contributing to frailty. Given the key role that the gut microbiota plays in important body functions, changes in this human microbial ecosystem may reduce colonization resistance and lead to immune changes resulting in increased disease risk. Therefore, the gut microbiota has often been considered a target to improve health in the senior population (Valdés *et al.*, 2017), together with the modulation of immune parameters, such as reducing the age-related pro-inflammatory status (Salazar *et al.*, 2017).

High-throughput techniques for the study of the microbiome

In the early era of microbiology, the study of the gut microbiota community commonly involved the isolation and cultivation of individual microbes from faeces; this allowed the discovery of coliforms by Teodor Escherich, who isolated the first gastrointestinal bacterium, *Bacterium coli commune*, later named *Escherichia coli* (Escherich, 1989). Following that, other major gastrointestinal bacterial members, such as *Bacteroides*, *Bifidobacterium* and *Bacillus*, were isolated (Rajilić-Stojanović and de Vos, 2014). However, efforts in this way were neglected by the fact that most of the microorganisms inhabiting in the gastrointestinal tract are obligate anaerobes and most of them were uncultivable. A significant advance in gut microbiota characterization was the introduction of the anaerobic culture techniques by Hungate (1969). Although cultivation-based approaches continue to be very important for the study of the gut microbiota, it was not until the emergence of the independent-culture techniques that the gut microbiota could be studied in detail without omitting uncultivable microbes. In addition, microbial taxonomy and phylogenetic relationships between organisms started to be established (Woese *et al.*, 1990).

DNA-based molecular techniques, such as denaturing gradient gel electrophoresis (DGGE), were used to identify gut dysbiosis associated with Crohn's disease, suggesting changes in *Clostridium* spp. and *Bacteroides* spp. (Martinez-Medina *et al.*, 2006; Scanlan *et al.*, 2006). This technique was also used to identify gut dysbiosis associated with cystic fibrosis, where lower abundance and temporal instability of bifidobacteria and *Clostridium* cluster XIVa in the faecal microbiota of patients with cystic fibrosis was observed (Duytschaever *et al.*, 2013). In a similar way, quantitative PCR (qPCR) was employed to compare the gut microbiota of healthy and hospitalized elderly, observing a reduction in the number of bacteria and the complete elimination of certain bacterial communities in hospitalized elderly individuals submitted to antibiotic treatment, as well as the proliferation of the opportunistic species *Enterococcus faecalis* (Bartosch *et al.*, 2004). Similarly, differences in gut microbiota composition were evidenced by this technique between a group of institutionalized elderly and a group of seniors (Salazar *et al.*, 2013). qPCR also evidenced a profound impact on the gut microbiota establishment of premature babies with respect to full-term neonates, with a lower number of anaerobes such as *Bifidobacterium* or *Bacteroides*, a higher number of pathogens such as *Klebsiella pneumoniae* and significantly reduced gut microbiota diversity occurring in preterm infants (Arboleya *et al.*, 2012). Finally, a study using the terminal-restriction fragment length polymorphism (T-RFLP) technique showed significantly reduced numbers of Bacteroidetes and a higher Firmicutes-to-Bacteroidetes ratio in obese subjects compared with non-obese subjects, which was consistent with data obtained by other methods (Kasai *et al.*, 2015). T-RFLP was used as well to analyse gut biopsy tissues in inflammatory bowel disease, showing an increment in species richness from control to ill but non-inflamed tissues, followed by a declined in fully inflamed tissues (Sepehri *et al.*, 2007).

Despite the fact that the association between microorganisms and host health was known previously to the advent of high-throughput sequencing techniques (Miller *et al.*, 1957), the emergence and fast progress of DNA sequencing technology has completely revolutionized human microbiota research in the last decade. These techniques are relatively fast and cost affordable and capable of generating billions of sequence reads, such as with Illumina HiSeq platforms, or providing long read lengths, such as with PacBio (Reuter *et al.*, 2015). This advance allows generating a big overview of the microbial members present in a sample, thus bypassing the need for culturing. Moreover, with the PCR-based sequencing approaches, enrichment of DNA templates with initial low concentration led to the identification of minority populations within complex communities. During the last decade, international research initiatives such as the Metagenomics of the Human Intestinal Tract consortium (MetaHIT) (Li *et al.*, 2014) or the Human Microbiome Project (HMP) (Human Microbiome Project Consortium, 2012) have taken advantage of those giant steps forward in the microbiome research and have generated many data and much knowledge. The last study from HMP has been published, updating body-wide metagenomics profile of the human microbiome, from the analyses of 1631 new samples, concluding that the Bacteroidetes to Firmicutes ratio should not be used as a defining attribute of an individual's gut microbiome (Lloyd-Price *et al.*, 2017). This rapidly moving advance in the sequencing platforms is producing a large number of data that far surpasses the ability to analyse them, creating in some cases a bottleneck in the interpretation of results. The bioinformatics field is growing fast as well and new tools are emerging to provide interpretation of data generated from NGS platforms.

Woese and Fox (1977) found that the ribosomal RNA (rRNA) gene could be used as marker gene to establish phylogenetic associations between organisms. Nowadays, the high-throughput sequencing of the small subunit rRNA gene is the most common method for gut microbiota characterization at compositional level. After total DNA isolation, PCR amplification with primers targeted towards highly conserved regions of the gene, 16S for prokaryotes and archaea and 18S rRNA for eukaryotes (including fungi) (Eckburg *et al.*, 2005; Amaral-Zettler *et al.*, 2009), generates a pool of amplicons derived from the different microbes present in a sample. Next, they are massively sequenced and clustered into operational taxonomic units (OTUs) by sequence similarity. This technique, also called 'marker gene survey', provides a broad census of the microbial species present in a sample. High-throughput amplicon sequencing has provided a large number of data on human microbiome diversity and composition, showing that Bacteroidetes, Firmicutes and Proteobacteria are the dominant taxa in the human gut (Eckburg *et al.*, 2005; Human Microbiome Project Consortium, 2012; Huse *et al.*, 2012; Yatsunenko *et al.*, 2012; Lloyd-Price *et al.*, 2017). The rRNA gene sequencing has also generated insight into host–microbe interactions, showing a direct correlation between gut microbiota, diet and health and yielding hypotheses about microbiota-based disease mechanisms (Turnbaugh *et al.*, 2009b; Muegge *et al.*, 2011; Claesson *et al.*, 2012). However, this technique presents some limitations, such as the biases that primer efficiency, PCR amplification and bioinformatics pipelines, among others, can introduce to results (Sharpton, 2014). The same principle, but focused on specific functional genes, was carried out to identify novel butyrate and propionate bacteria producers (Walker *et al.*, 2014; Louis *et al.*, 2010; Reichardt *et al.*, 2014). In other studies, amplifications of the RNA polymerase β-subunit (*rpoB*) gene were performed for the differentiation of *Bifidobacterium* species in infant samples (Kim *et al.*, 2010; Murphy *et al.*, 2015).

Cataloguing members of the gut microbiota has developed into routine; however, the foundational understanding of the functionality of each individual member in their ecosystem still remains a challenge. With this purpose, meta-omics technics focused not only on DNA, but also on RNA, proteins or small molecules, with these providing huge progress in the comprehension of the intricate crosstalk among gut bugs and gut bugs and the host.

Shotgun metagenomics sequencing extends the information provided by 16S/18S rRNA amplicon sequencing to study the entire genomic content of all microbes, cultured or uncultured, present in a sample. This approach avoids previous sequencing limitations as all the genomic DNA is trimmed into tiny fragments that are all individually sequenced, instead of sequencing only a specific amplicon. It will provide both taxonomic and biological function information encoded by the genome (Sharpton, 2014). This technique has evolved to address the questions of 'who is present?' and 'what they can do?' (Oulas *et al.*, 2015). Moreover, it allows the identification of bacteriophages, members of the gut virome, which modulate the bacterial community and interact with the host, playing an important role in host health (Reyes *et al.*, 2010, 2015). There are, however, some important limitations to the use of shotgun metagenomics such as its higher cost, complexity and difficulty of data analyses, as well as the unwanted host DNA that usually contain (Prakash and Taylor, 2012; Sharpton, 2014). The first human intestinal metagenome study, dating from 2006 (Gill *et al.*, 2006) and carried out with samples from two healthy adult volunteers, revealed an enrichment in metabolic pathways related to carbohydrates, amino acids and xenobiotics. Modifications in the gut microbiota and its link with human metabolism on ageing have been corroborated with this technique in a study carried out by Rampelli *et al.* (2013). A loss of genes for SCFA production and an overall decrease of the saccharolytic potential in the intestinal metagenome of elderly with respect to younger adults were observed. Shotgun sequencing has been used to define functional profiles of gut microbiota in the context of several diseases (Qin *et al.*, 2010; Morgan *et al.*, 2012; Zeller *et al.*, 2014).

Whereas metagenomics focuses on all of the functions present in a community (the functional *potential*), metatranscriptomics describes the functional activity of the gut microbiota expressed at a given time and depending on the specific environmental conditions: 'what they are doing?'. This technique involves isolation of RNA to be converted to a cDNA library, which is subsequently sequenced in a high-throughput platform. Next, reads are mapped to a reference genome or transcripts, or assembled *de novo* (Wang *et al.*, 2009). Metatranscriptome data indicated that the gut microbiota contributes to reinforce and increase the metabolic potential of the host, as almost 50% of genes expressed by this microbial community are involved in transport and metabolism of carbohydrates, amino acids, lipids, nucleic acids and energy production (Gosalbes *et al.*, 2011), which corroborates findings obtained with other omics techniques (Gill *et al.*, 2006). Metatranscriptomics have been used to detect microbial metabolic perturbations and to show the influence of environmental factors on gut microbiome functions, evidencing differences between the functional potential and the activity of the gut microbiome (Franzosa *et al.*, 2014). It also allowed the analysis of the gut virome (Santiago-Rodriguez *et al.*, 2015) and the study of microbiota–host interactions (Camp *et al.*, 2014). Moreover, *Fusobacterium nucleatum* has been discovered to be involved in colorectal cancer by this technique (Castellarin *et al.*, 2012). In spite of this, metatranscriptomic usually becomes more challenging than metagenomics because of the need for conversion of mRNA to cDNA, the lack of reference genomes or databases available and the short-half-life of mRNA molecules (Walker, 2016).

Metaproteomics, defined as the large-scale profiling of the complement of proteins produced by a complex microbial ecosystem (Wilmes and Bond, 2009), presents some advantages over metatranscriptomics because the study of proteins provides a more representative overview of the final functional activity of the community (Walker, 2016). Despite its limitations (Walker, 2016), this techniques has been largely employed to characterize functional roles of gut microbiota in health and disease conditions (Lee *et al.*, 2017).

Together with metaproteomics, metabolomics closes the entire pathway from genes to the functionality of the gut microbiota. The study of the final products (metabolites) of bacterial metabolism present in a sample provides a better knowledge of the functionality and physiology (Ursell *et al.*, 2014). Metabolomics involves the identification and quantification of the metabolites (Fiehn, 2002) and it can be performed in a targeted or untargeted way. The targeted method is based on the analysis of different classes of previously defined compounds, such as amino acids, lipids, carbohydrates or fatty acids, whereas the non-targeted approach provides a global profile picture of the metabolic diversity of the sample (Smirnov *et al.*, 2016). Different methodology and analytical platforms are employed in metabolomics, including gas chromatography (GC) or high performance liquid chromatography (HPLC) coupled to mass spectrometry (MS) or proton nuclear magnetic resonance spectroscopy (H-NMR), among others (Vernocchi *et al.*, 2016). The use and combination of more than one methodology is necessary to analyse the complete metabolome of a sample, because of the diversity in properties and concentrations of the metabolites (Smirnov *et al.*, 2016). Some limitations are linked to this technique, such as the lack of good and complete reference databases, the resolution limits or the difficulty of assigning the production of particular metabolites to specific microbes in an accurate way (Walker, 2016). The number of metabolomic studies in gut microbiota have increased in the last years (Smirnov *et al.*, 2016), evidencing a gut microbiota modulation related to diet and ageing. De Filippis *et al.* (2016) analysed the effects of a Mediterranean diet on the human gut metabolome profile and the effects of synbiotic foods on gut metabolome were studied by Vitali *et al.* (2010). Age-related metabolic changes were found in urine and faeces from mice (Calvani *et al.*, 2014), as well as in Italian centenarians, by using an H-NMR approach (Collino *et al.*, 2013).

The high-throughput techniques for the gut microbiome study mentioned above have their 'pros' (advantages) and 'cons' (limitations). The integration of the information generated by all of them could provide an extensive and more comprehensive snapshot of the gut microbiota and its relationship with the host, in addition to a definition of biomarkers and interventional targets in different conditions and for different populations (Abram, 2015; Zhang and Zhao, 2016). Despite this, to determine which particular bacteria has a specific function, or how its function is affecting the host, the research still has to come back to the 'the microscope' (Troponi *et al.*, 2017) and the culturing approach. In the last years, Raoult's research group, through a high-throughput culturing approach, so-called 'culturomics', has 'broken the records' and increased the number of species isolated and cultured from the human gastrointestinal tract to over 1000 (Lagier *et al.*, 2012, 2015, 2016). They were able to cultivate organisms corresponding to sequences previously not assigned by metagenomics. Culturomics combines high-throughput culture conditions and matrix-assisted laser desorption/ionization-time of flight (MALDI-TODF) or 16S rRNA sequencing for the identification of growing colonies (Lagier *et al.*, 2016). This rebirth technique improves the capture of gut diversity, improving our understanding of the gut environment.

The evolution of the intestinal microbiota through ageing

Traditionally, the establishment of the intestinal microbiota and the microbial colonization of the gastrointestinal tract have been considered to occur during and after delivery; however, the presence of microorganisms in placenta and amniotic fluid (Collado *et al.*, 2016) suggests a primary fetal colonization. The classic pattern of colonization at the beginning of life involves mainly facultative anaerobes, which are replaced over time by strict anaerobic microorganisms due to the consumption of oxygen by the first colonizers (Fanaro *et al.*, 2003). The initial gut microbiota composition is simple, dynamic and very unstable and undergoes marked fluctuations. Some of the most influential factors on the nascent microbiota appear to be the mode of delivery, feeding practices, maternal weight or lifestyle and the use of antibiotics (Nogacka *et al.*, 2017, 2018). Accumulating evidence indicates that the establishment and development of the microbiota in early life and its cross-talk with the host is especially critical in the developmental programming of adult health and disease. Moreover, there is increasing concern about the potential long-term effects of microbiota alterations in the neonatal early life. This early microbiota seems to provide a stimulus required for an adequate development of the gut and immune functions (Nogacka *et al.*, 2018). The microbiota development process determines to some extent the predisposition to develop diseases in early and later life including metabolic and immune pathologies such as necrotizing enterocolitis, obesity and type 2 diabetes, chronic intestinal inflammation, or allergy, among others (Goulet, 2015).

The classical microbiota development begins from low diversity and complexity and evolves until reaching a complex structure that stabilizes at 2–3 years of age (Yatsunenko *et al.*, 2012). The adult-like intestinal microbiota is characterized by a set of microorganisms that mainly inhabit the gastrointestinal tract, and have co-evolved with humans in a commensal way, reaching its highest concentration in the colon (about 10^{14} cells) (Arumugam *et al.*, 2011). A ratio between intestinal bacteria and total human cells was estimated as 10 : 1, although considering only the nucleated human cells (excluding erythrocytes) the approximate ratio is 1.3 : 1 (Sender *et al.*, 2016). The composition of adult intestinal microbiota at the phylum level is represented almost entirely by two major phyla, Bacteroidetes and Firmicutes. The ratio between them is frequently used as an indicator of alterations and bacterial shifts, although, based on data generated in the HMP project about the human microbiome, researchers conclude that the Bacteroidetes-to-Firmicutes ratio should not be used as a defining attribute of an individual's gut microbiome (Lloyd-Price *et al.*, 2017). The remaining minority phyla of the human microbiota in descending order are Actinobacteria, Proteobacteria, Verrucomicrobia and Euryarchaeota (Arumugam *et al.*, 2011).

The microbiome is usually recognized as our second genome (Aagaard *et al.*, 2013) and contributes extensively to our physiology and metabolism (Qin *et al.*, 2010). There are some well-established functions attributed to the intestinal microbiota, among which the following are noteworthy: (i) the maintenance of the intestinal epithelium integrity and functionality, contributing to enhance the intestinal barrier, to nutrients supply and to the protection against pathogens, (ii) inhibition of pathogen adhesion to intestinal surfaces,

(iii) modulation and maturation of the immune system, (iv) degradation of non-digestible carbon sources and (v) production of different metabolites such as vitamins and SCFA (Tojo *et al.*, 2014). The intestinal microbiota is relatively stable throughout adulthood, although, as a dynamic ecosystem, it moderately fluctuates and is susceptible to some variations owing to stress, antibiotics, diet and lifestyles (Rodríguez *et al.*, 2015).

In the elderly, an inverse process occurs that resembles a mirror image of the neonatal gut colonization and the microbiota becomes unstable again and dysbiosis frequently appears (Salazar *et al.*, 2017). Age-related changes in the gut microbiota are associated with physiological changes in the gastrointestinal tract, as well as in dietary patterns, with a concomitant decline in the normal function of the immune system that may contribute to increased risk of infection and frailty (Claesson *et al.*, 2011, 2012; Odamaki *et al.*, 2016). An imbalanced diet frequently occurs because of problems in the smell sense, accompanied by difficulties in swallowing and masticatory dysfunction. In addition, the reduced intestinal motility leads to faecal impaction and constipation and increases the intestinal transit time. Besides, a slower intestinal transit times leads to altered nutrient availability and absorption, reduced stability and diversity of microbial communities, thinning of the mucosal lining and subsequent dysfunction of the intestinal barrier, and increased inflammation, all of them being common manifestations of ageing of the gastrointestinal tract (Biagi *et al.*, 2012). These factors inevitably affect the functionality of the microbiota and are linked with alterations of microbiota composition and metabolism.

There is no currently agreed definition of old age-specific gut microbiota profile due to the high inter-individual variability, differences in diet and lifestyle and the unclear definition of the term 'elderly'. In addition, there are a limited number of studies in the literature that describe the composition of the microbiota of elderly individuals, which restricts our ability to establish cause and effects relationships. Briefly, the elderly's gut microbiota is characterized by a reduced bacterial diversity, shifts in the dominant species, a decline in beneficial microorganisms, increase of facultative anaerobic bacteria, decreased levels of total SCFA and changes in the ratios among the different SCFA with respect to that found in younger adults (Claesson *et al.*, 2012; Salazar *et al.*, 2013, 2014). Opportunistic pathogenes also appear more frequently at old age (O'Toole and Claesson, 2010; Kumar *et al.*, 2016; Salazar *et al.* 2017). Moreover, the microbiota composition is also influenced by the well-established driving force, which is the geographical factor (Yatsunenko *et al.*, 2012). Therefore, there are numerous confounding data about the changes due to age in several microbial groups. Thus, for example, a comparative study of the microbial composition in four European countries indicated age-related structural differences that differ from country to country (Mueller *et al.*, 2006). Nevertheless, some trends are repetitively observed such as the decrease in bifidobacteria in the elderly population, confirmed by several studies using different methodologies (Gavini *et al.*, 2001; Hopkins *et al.*, 2001; Biagi *et al.*, 2012). The gut microbiota analysis in old population showed a clear shift to a *Clostridium* cluster IV-dominated community in the elderly Irish population with respect to younger adults (Claesson *et al.*, 2011) but this observation was the inverse in Austrian elders (Zwielehner *et al.*, 2009) and in a Finnish population (Mäkivuokko *et al.*, 2010). This fact may be due to the inclusion of the genus *Faecalibacterium*, which shows a country-specific pattern, with higher levels observed in French, German, Swedish (Mueller *et al.*, 2006) and Irish (Claesson *et al.*, 2011) populations and lower levels in Italian (Mueller *et al.*, 2006; Biagi *et al.*, 2016), Spanish (Salazar *et al.*, 2013) and Chinese (Wang *et al.*, 2015) cohorts. *Clostridium* cluster

XIVa has also been found to be associated with a country–age relationship. In German seniors increased populations have been observed (Mueller *et al.*, 2006) whereas decreased populations in Spanish (Salazar *et al.*, 2013), Italian (Mueller *et al.*, 2006; Biagi *et al.*, 2016) and Japanese (Hayashi *et al.*, 2003) elders, as compared with their respective adult cohorts, have been reported for these bacterial groups. Similar country-age discordances occur with *Bacteroides – Prevotella* group. That cluster is under-represented in Italian (Mueller *et al.*, 2006; Claesson *et al.*, 2011; Biagi *et al.*, 2016), English (Woodmansey *et al.*, 2004) and Spanish (Salazar *et al.*, 2013) cohorts and is over-represented in German, Swedish (Mueller *et al.*, 2006), Austrian (Zwielehner *et al.*, 2009), Finnish (Mäkivuokko *et al.*, 2010), Irish (Claesson *et al.*, 2011) and Japanese (Odamaki *et al.*, 2016) cohorts. Nevertheless, in spite of discrepancies existing among studies in different countries, there is a rather general agreement that the variations in the gut microbial composition at senescence frequently involve lower levels of *Blautia coccoides* (also known as *Clostridium* XIVa group), *Bifidobacterium* and *Faecalibacterium* groups as compared with adults (Hopkins *et al.*, 2001; Biagi *et al.*, 2010; Claesson *et al.*, 2011, 2012; Salazar *et al.*, 2013, 2017; Odamaki *et al.*, 2016). Other relevant intestinal microbial populations such as the phylum Bacteroidetes and the *Lactobacillus* group remain controversial; although some authors have reported lower levels in aged subjects (Hopkins *et al.*, 2001; Hayashi *et al.*, 2003; Bartosch *et al.*, 2004; Woodmansey, 2007; Biagi *et al.*, 2010; Tiihonen *et al.*, 2010; Claesson *et al.*, 2011; Salazar *et al.*, 2013), others have observed increases at older ages (Mäkivuokko *et al.*, 2010; Tiihonen *et al.*, 2010; Salazar *et al.*, 2013; Odamaki *et al.*, 2016). However, it is still unclear whether the variability in the results obtained for certain microbial groups is related to actual population differences or to methodological issues, such as the different techniques used for determining microbial abundancies. As expected from the differences in microbiota observed between elderly and younger adults, the production of bacterial metabolites, mainly SCFA, is also altered at old age.

Recent studies comparing the trajectory of the microbiota from adults to elderly, centenarians and long-lived semi-supercentenarians (105–110 yeasr) evidenced a cumulative abundance decreasing of some symbiotic bacterial taxa (mainly belonging to the Ruminococcaceae, Lachnospiraceae and Bacteroidaceae families) with age. In extreme longevity, the microbial diversity increases again: facultative anaerobes and allochthonous opportunistic bacteria occur, but at the same time some new microorganisms, not found in younger adults and elder, seem to emerge, whereas a higher prevalence than in younger individuals occur for some health-associated bacterial groups (e.g. *Akkermansia*, *Bifidobacterium* and Christensenellaceae) (Biagi *et al.*, 2016). Regarding the Firmicutes to Bacteroidetes ratio, frequently used in the scientific literature as an indicator of bacterial shifts, variable and controversial results have been reported (Salazar *et al.*, 2017). However, it is not currently possible to determine if these features observed in extremely old people were already present in the same individuals at a younger age and/or if they are associated with their past lifestyle (Biagi *et al.*, 2016).

Gastrointestinal infections in the elderly

In addition to structural and functional changes, increase in oxidative stress and inflammation of the intestinal mucosa naturally occurring in senescence (Soenen *et al.*, 2016), other factors such as antibiotic and non-steroidal anti-inflammatory drug treatments, changes in diet and/or institutionalization contribute to pathological alterations of the intestinal microbiota (Claesson *et al.*, 2012; Tarnawski *et al.*, 2014). *Clostridium difficile* is the most frequent pathogen in bacterial infections associated with health care and ageing (Magill *et al.*, 2014), followed by *Helicobacter pylori* and, to a lesser proportion and depending on the cohort, *Bacteroides fragilis*, *Klebsiella oxytoca* and enteropathogenic *Escherichia coli* (Magill *et al.*, 2014; Kullin *et al.*, 2015; Luo *et al.*, 2017).

C. difficile is an anaerobe, spore-forming, Gram-positive bacterium found in the intestinal tracts of both animals and humans (Elliott *et al.*, 2017). Institutionalization is one of the main risk factor for *C. difficile* infection (CDI), this fact contributing to increases up to 21% of carriage rate in individuals under short- or long-term care in hospitals, in comparison with 1.6% in the community (Rea *et al.*, 2012). It has been indicated that *C. difficile* can be acquired by patients with diarrhoea caused by other microorganisms after some days of stay in hospital (Rand *et al.*, 2015). Antibiotic treatment is another main factor that facilitates infection by this pathogen. In fact, different antibiotic treatments lead to different intestinal microbiota modifications and, as a result, different susceptibilities to CDI (Willing *et al.*, 2011; Lewis *et al.*, 2015). A decrease in bacterial diversity occurs during CDI in elderly patients, with health-promoting bacteria including *Faecalibacterium* and *Bifidobacterium* being under-represented in those patients (Rea *et al.*, 2012). A recent study comparing CDI patients with healthy controls that were not under antibiotic treatment revealed a decrease in alternative health-related bacteria, as *Akkermansia*, and in some groups of commensal bacteria, as members of the families Ruminococcaceae and Lachnospiraceae, along with members of the genera *Bacteroides* and *Alistipes* (Milani *et al.*, 2016). In the two last studies mentioned above, authors found an increase in some recognized or opportunistic pathogens, such as *Enterococcus*, *Staphylococcus*, *Klebsiella*, *Escherichia* or *Helicobacter* members, among others (Milani *et al.*, 2016). Residents in nursing houses with CDI had more comorbidities than non-CDI residents, including diabetes, urinary incontinence or chronic obstructive pulmonary disease (Zarowitz *et al.*, 2015). Additionally, residents with CDI acquired before institutionalization were less likely to carry a burden of illness than residents who acquired CDI in the nursing house (Zarowitz *et al.*, 2015). Common antibiotic treatments against CDI, including vancomycin, metronidazole and fidaxomicin, show a relatively high recurrence rate (Kociolek and Gerding, 2016); however, promising results are being obtained with the novel therapy consisting of faecal microbiota transplantation, which reaches an 83–94% response rate in recurrent CDI (Bagdasarian *et al.*, 2015). As comorbidities frequently occur in elderly people suffering CDI, faecal microbiota transplantation should be considered case by case; experts stress the importance of prevention and combination of antimicrobial and non-antimicrobial complementary treatments as first-line options in CDI (Asempa and Nicolau, 2017).

H. pylori is a microaerophilic, Gram-negative, spiral-shaped and flagellated bacterium found in the gastrointestinal tract (Tomb *et al.*, 1997). *Helicobacter* infections are more frequent as the population gets older, starting to rise at the age of 40–55 years and keeping the same levels of prevalence in the elderly (Genta *et al.*, 2017). *H. pylori* infection (HpI) can lead to Barrett's oesophagus, peptic ulcers, intestinal metaplasia and stomach cancer (Reva *et al.*, 2015). Some studies have related the presence of *H. pylori* in the elderly with changes in plasma levels of ghrelin and in the ghrelin-to-obestatin ratio, which are known to play an important role in appetite regulation (Gao *et al.*, 2009a; Kim *et al.*, 2017). In particular, Kim *et al.* (2017) found that the reduction of ghrelin levels in plasma, and not the presence of *H. pylori per se*, were related to the intestinal metaplasia associated with infection by this bacterium. As one of the main risk factors of infections in the elderly is malnutrition, the impairment in ghrelin production during HpI can contribute to a reduction in appetite and food intake, thus worsening even more the state of malnutrition and increasing the risk of subsequent infections (Soenen *et al.*, 2016). The eradication of *H. pylori* contributes to improve several symptoms associated with HpI, such as dyspepsia or peptic ulcers. In those areas with high levels of resistance to the preferentially used antibiotics clarithromycin and metronidazole, the treatment of HpI with bismuth quadruple therapy is recommended (Malfertheiner *et al.*, 2017).

Norovirus is a single-stranded RNA virus (Zheng *et al.*, 2006) that, together with *C. difficile*, was the cause of most of the gastroenteritis deaths that occurred during 1999–2007 (83%) in the US elderly population (Hall *et al.*, 2012). The most common population hospitalized because of norovirus infection are children and the elderly, with infection being more severe in the former group (Yi *et al.*, 2016). Treatment with proton pump inhibitors, a very common drug used among elderly to alleviate digestive discomfort, has been correlated with higher rates of norovirus infection (Prag *et al.*, 2017). Norovirus are clustered in seven genogroups and the majority of the outbreaks in humans have been caused by the genotype GII.4. Although some vaccines are under trial, the duration of protection after administration is still undetermined (Aliabadi *et al.*, 2015).

An integrated approach to changes on the intestinal microbiota, dietary patterns and lifestyle factors: from adulthood to senescence

The first evidence of the influence of food and exercise in health appears in the Hippocratic corpus in the fifth century bc. Currently, there is solid scientific evidence about the importance of adequate nutritional status for the maintenance of good health in all stages of life, and particularly among the elderly, in order to prevent the appearance and progression of several chronic diseases. Nutritional requirements vary throughout life as a consequence of the modification of a plethora of physiological parameters such as body composition, basal metabolic rate or nutrient absorption, which take place together with other factors very common in the elderly, such as a sedentary lifestyle, loneliness or depression. Most of these age-related changes are intrinsically linked to the functioning of the digestive system and to the modifications of dietary patterns that occurs from adulthood to senescence. For example, the appearance of dysgeusia and xerostomia or the deterioration in dentition may impact the appetite in many ways and alter food choices, which can lead long term to malnutrition (Clemente *et al.*, 2012; Amarya *et al.*, 2015) and greatly influence the gut microbial composition, diversity and functionality (Flint, 2012). As gut microorganisms are pivotal for homeostasis in the intestine, the regulation of the immune system (Garrett *et al.*, 2010; Li *et al.*, 2016) and the prevention of major chronic non-communicable inflammatory and metabolic disorders (Shanahan *et al.*, 2017), diet could be one of the environmental factors with a greater impact on the future health of the elderly population.

Health implications of changes on the intestinal microbiota through age

Sorting out the type of microbiota is the first step in the design of functional foods specifically focused on the elderly. Nevertheless, this is a difficult task considering the high inter-individual variability existing in the intestinal microbial composition and the pool of factors that influence its establishment and composition along life. Different studies have reported changes in the gut microbial composition at senescence (see 'The evolution of the intestinal microbiota through ageing'). We have analysed differences in the nutritional status, intestinal microbial composition and intestinal microbial metabolites in a sample cohort of 114 volunteers from northern Spain, with non-declared pathology, who were stratified by age in three groups: adults (19–55 years), seniors (56–65 years) and elderly (65–95 years) (Table 1.1). Similarly to those reported by other authors, we have also found some differences in the faecal microbial composition related to age (Table 1.2). In our cohort, individuals older than 65 years showed lower levels of *Bacteroides*, *Bifidobacterium*, *Bl. coccoides* and *Faecalibacterium* together with a lower faecal concentration of the major SCFAs. The loss of the adult-associated microbiota and a reduced diversity in elderly subjects have been repeatedly reported to be associated with frailty (Claesson *et al.*, 2011; Jackson *et al.*, 2016) and have been related to a pro-inflammatory status. This pro-inflammatory status could be a possible driver of the ageing process (Biagi *et al.*, 2010), contributing to increased frailty and comorbidities and with the nutritional status (Lynch *et al.*, 2014).

There is not a specific age that determines the changes from adult to an elderly microbiota; this transition usually occurs gradually, is dependent on external factors and is different from one individual to another. It seems therefore that the maintenance of a healthy microbiota could have benefits to maintain a good quality of life associated with old age. In general, a high microbial diversity is important for intestinal homeostasis, favouring the tolerance of the microbial environment to the modification in external factors such as dietary ones (Turnbaugh *et al.*, 2009b; Lozupone *et al.*, 2012). Several studies have reported that some members of the *Bl. coccoides* group (belonging to *Clostridia* cluster XIVa) can impact human health by diverse mechanisms including the production of butyrate, which

Table 1.1 General characteristics of the study sample (northern Spanish healthy population) by age group

Characteristic	Elderly ($n=40$)	Senior ($n=37$)	Adult ($n=37$)
Age (years)	82 ± 6*	61 ± 3*	41 ± 8*
BMI (kg/m^2)	29.15 ± 4.14	25.91 ± 3.39*	29.42 ± 4.51
Energy intake (kcal/day)	1687.03 ± 393.97*	1929.93 ± 562.10	1867.32 ± 510.20
Protein intake (g/day)	82.54 ± 20.59	94.53 ± 29.33*	77.25 ± 28.09
Lipid intake (g/day)	73.68 ± 21.34	77.14 ± 22.67	73.10 ± 26.03
Carbohydrate intake (g/day)	171.61 ± 41.76	198.97 ± 83.86	220.48 ± 59.22*
Fibre intake (g/day)	15.96 ± 5.06	23.02 ± 9.14*	17.84 ± 5.37

*Significant differences ($P < 0.05$).
The results are presented as the mean ± standard deviation. Statistical analyses among groups were carried out using an analysis of variance (ANOVA) test.

Table 1.2 Levels of the major intestinal microbial groups (log no. of cells per gram of faeces) obtained by qPCR and short-chain fatty acids (mM) determined by gas chromatography

	Elderly ($n=40$)	Senior ($n=37$)	Adult ($n=37$)
Microbial group:			
Akkermansia	6.55±2.84a	6.26±2.12a	5.26±1.43a
Bacteroides	7.78±2.82a	9.31±0.80b	7.93±1.06a
Bifidobacterium	6.85±2.64a	7.84±1.47a,b	8.38±0.95b
Bl. coccoides	5.54±2.44a	7.51±1.56b	8.94±0.58c
Lactobacillus	6.40±2.80a	5.85±1.27a	6.32±0.99a
Faecalibacterium	6.14±1.64a	7.07±0.74b	8.08±0.90c
SCFA:			
Acetic acid	20.97±13.41a	30.47±9.88b	51.22±16.49c
Propionic acid	8.05±6.15a	13.59±6.10b	15.73±7.71b
Butyric acid	7.03±6.78a	12.39±9.46b	9.48±3.77a,b
Total SCFAs (acetic, propionic and butyric acid)	36.06±25.37a	56.44±24.10b	76.44±25.4

The results are presented as mean±deviation. Different letters (a, b, c) indicate significant differences among age group ($P<0.05$) (statistical analysis: ANOVA and least significant difference *post hoc* test).

influences host carbohydrates and lipids metabolism (Louis and Flint, 2009; Rios-Covian *et al.*, 2016), favouring the production of secondary bile acids (Kitahara *et al.*, 2000) and inducing formation of T-regulatory cells (Atarashi *et al.*, 2013). In addition, *F. prausnitzii* is one of the most abundant butyrate producers in the human colon and has been proposed to have anti-inflammatory properties (Chun *et al.*, 2007).

Consequently to microbiota alterations in the elderly, the faecal SCFA profile is also modified. The concentrations of SCFA in the colon are critically important for immunoregulation (Maslowski and Mackay, 2011) and for maintaining gut and overall health (Gao *et al.*, 2009b; Peng *et al.*, 2009; Fukuda *et al.*, 2011). Therefore, the lower excretion of all the evaluated SCFA in our elderly Spanish cohort is in good agreement with that previously reported in the literature (Claesson *et al.*, 2011; Salazar *et al.*, 2013) and may also be related to imbalances in host health.

Dietary patterns as related to age and microbiota

It has been estimated that between 2% and 16% of institutionalized elderly people present an inadequate intake of protein and calories (Smeedings, 2001), which could increase the risk of malnutrition and exert an important impact on gut microbiota. In this way, it has been demonstrated that in adults the major microbial 'enterotypes' (Arumugan *et al.*, 2011) were clearly determined by the long-term dietary patterns of an individual: although diets with a high content of protein and animal fat were associated with a *Bacteroides* enterotype, those with a higher proportion of carbohydrates were dominated by *Prevotella* (Wu *et al.*, 2011). The sample cohort from northern Spain that we have examined in the present chapter presents a moderate reduction in the calorie intake with advancing age, linked to a decrease in the carbohydrates and proteins from diet. It was, however, enough to satisfy the requirements of the sample, taking into account the reduction of the basal metabolic rate inherent to the ageing process. In addition, protein intake accounts for approximately 19% of the total energy, satisfying the standard recommendations. Protein under-nutrition usually arises associated with weight loss, which does not happen in our sample whose body mass index (BMI) is above 29 kg/m^2. This value equates to an overweight BMI category (Salas-Salvado *et al.*, 2007), so not pointing to significant deficiencies in the intake of macronutrients in the elderly group from our cohort.

The maintenance of a high dietary variety is essential for an overall optimal nutritional status and health outcomes in frail elderly people (Bernstein *et al.*, 2002). In spite of this, foods with low palatability, such as vegetables and fruits, and those difficult to chew or to digest, like meats, are often excluded from the diet or are reduced with ageing. Based on this premise, changes that take place drastically with age could have a major impact on the intestinal microbial ecosystem. To examine whether there is a change in the dietary pattern in our sample associated with age we analysed the intake of the major food groups in our sample according to age. Significant differences in the intake of sugar and sweets, alcoholic and non-alcoholic beverages, cereals, fruits, eggs, milk and milk products, fish and seafoods, roots and vegetables were observed according to age (Fig. 1.1). Interestingly, our data show a striking reduction in the intake of cereals, fruits and vegetables at the age of 65 years. Thus, despite the absence of a consensus about a cut-off point to define the beginning of old age, it seems possible that, from a nutritional point of view, the age of 65 years marks a change in the feeding pattern.

As we commented previously, foods may not only provide the energy and structural components, but also contain carbon and nitrogen sources that are essential elements for the growing and metabolism of the intestinal microbiota (Li *et al.*, 2017), and bioactive compounds with a demonstrated effect as microbiota modulators. From them, fibres, defined by the Association of Cereal Chemists as 'the edible parts of plants or analogous

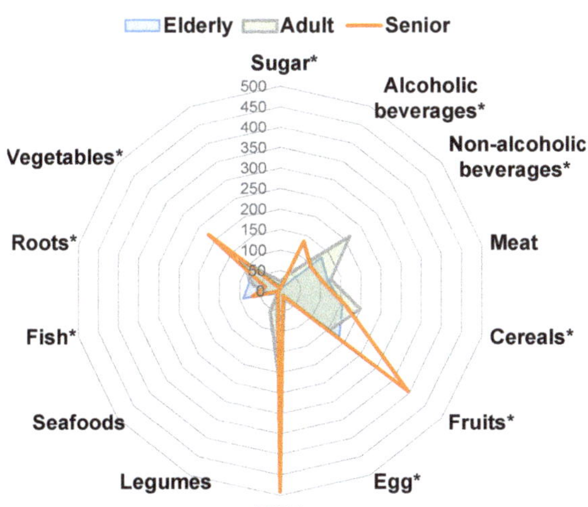

Figure 1.1 Mean intake of major food groups (grams/day) according to age.

carbohydrates that are resistant to digestion and absorption in the human small intestine with complete or partial fermentation in the large intestine' (AACC, 2001), seem to be a key dietary component in the general population with a special focus on aged people. It has been reported that only about 40% of subjects above 71 years have dietary intakes above the adequate level of this dietary component (IOM, 2010); in line with these data, the intake in our sample is far from the 25–30 g/day recommended for protecting against heart disease, the prevention of several metabolic conditions and the proper maintenance of intestinal health (IOM, 2010). The reduction of carbohydrates accessible to gut microbes from fibre-containing foodstuffs may result in a long-term reduction of microbiota diversity and in the appearance of 'unhealthy' microbiomes (Sonnenburg *et al.*, 2005, 2016; Yatsunenko *et al.*, 2012). In this regard, some types of fibres, such as inulin, fructo-oligosaccharides or resistant starch, have shown a prebiotic effect promoting the grow of colonic bifidobacterial (Yen *et al.*, 2011; Petry *et al.*, 2012; Yang *et al.*, 2013).

Fruits, vegetables and cereals are food groups containing different mixtures of fibres and phytochemicals that may have shown a positive modulatory effect on some intestinal bacterial populations (Lee *et al.*, 2006; Vendrame *et al.*, 2011; Massot-Cladera *et al.*, 2012; Queipo-Ortuño *et al.*, 2012; Guglielmetti *et al.*, 2013), contributing to the maintenance of a well-balanced microbiota in the elderly (Gutiérrez-Díaz *et al.*, 2017; Salazar *et al.*, 2017). Specifically, there is recent evidence on a positive association between the frequency of consumption of fruits and vegetables and some intestinal microorganisms such as *Lactobacillus*, *Bl. coccoides* and *Prevotella* (La-ongkham *et al.*, 2015). Although it is not possible to establish causal relationships, vegetables and fruits were the major dietary sources of pectins in a population of Spanish healthy adults (Cuervo *et al.*, 2015). Pectins can be fermented by certain colonic bacteria such as *Bacteroides* resulting in the production of SCFA (Topping and Clifton, 2001), which contributes to increase the production of SCFA in the colon. Besides, fruit and vegetables contain sources of vitamin C and carotenoids, compounds that have known anti-inflammatory activity and also provide other bioactive compounds, such as polyphenols, that are involved in the regulation of some metabolic conditions linked to inflammation and oxidative stress (Shen *et al.*, 2012). Thus, the identification of novel nutritional and biological biomarkers of ageing will be a promising target in future ageing research.

Restoring a balanced microbiome through diet and lifestyle

The increased susceptibility to disease in the elderly, which is frequently related to malnutrition, a pro-inflammatory status of the immune system and the impaired intestinal microbiota functionality, provides the rationale for developing dietary strategies targeting the intestinal microbiota, the immune system and the specific nutritional needs in old age.

Nutritional needs in the elderly can differ from those of middle-aged adults (McCarty and DiNicolantonio, 2015). Although there is not a consensus on the nutritional deficiencies in the elderly and these can vary from one human group to another, some nutrients have been linked to a better immune response: monounsaturated fatty acids (MUFA), beta-carotene, vitamins A, B_6, C and D, and bioactive compounds. In addition, in the elderly some macro- and micro-nutrients fall frequently below the recommended intake levels (Trumbo *et al.*, 2002): proteins, fibre, iron, B group vitamins (including folic acid) and vitamin D (Salazar *et al.*, 2017).

It is currently clear that the impact of diet on the microbiota is more important in the elderly than in young adults (Claesson *et al.*, 2012; Jeffery *et al.*, 2016). It is worth remarking that not all changes occurring in the microbiota as age advances may necessarily have a detrimental health effect, as functional redundancy exists in this microbial ecosystem. However, the reduction in microbial diversity, the shift in dominant species and the reduced levels of some groups of bacteria such as those producing butyrate, the increase of facultative anaerobic bacteria, the decreased levels and imbalanced production in the gut of SCFA, and increased levels of lactate, methane and branched-chain fatty acids are generally considered relevant targets for intervention (Salazar *et al.*, 2017). In this respect, it seems that in the elderly the loss of the community-associated microbiota, considered as the group of microorganisms shared by the individuals from a given social group, has been correlated with a decrease in microbial diversity and with increased frailty (Claesson *et al.*, 2011; Jackson *et al.*, 2016; Jeffery *et al.*, 2016).

Intervention targets with regard to the intestinal microbiota and the immune system may differ among elderly groups; in fact, nutritional profiles and the characteristics of the intestinal microbiota and the immune system can vary depending on the geographical location and dietary and social patterns (Mueller *et al.*, 2006; Turnbaugh *et al.*, 2009a; Claesson 2012). On the other hand, there is not a specific age determining the shift from an adult to an elderly microbiota but it depends on the external factors just commented, and possibly some others (O'Toole and Jeffery, 2018). Therefore, prior to designing nutritional interventions it is necessary to identify the specific action targets for specific human groups. The microbiota of healthy subjects is considered the reference for comparison, in order to define targets for restoring a balanced microbiota. However, a 'healthy microbiota profile' cannot be defined 'a priori'. Thus, human groups of reference must have a socio-economic status (geographical location, social habits, historical past, dietary habits) as close as possible to that of the population under study. This is especially important in the elderly in whom prolonged exposure to different environments could make it difficult to identify variations specifically due to ageing and those due to other factors. The study of centenarians and super-centenarians and their offspring deserves special mention, as a model to disentangle genetic and environmental lifestyles that determine healthy ageing and the characteristics of their microbiota in comparison to septuagenarians and octogenarians (Santoro *et al.*, 2018). In long-term planning, longitudinal studies along years of the intestinal microbiota, dietary patterns, lifestyle factors and health status over ample cohorts will provide the most complete information about changes in the intestinal microbiota with ageing, the influence of external factors, such as diet, on these changes and, ultimately, how these variables determine healthy ageing.

A moderate enhancement of the colonic fermentation of dietary fibre could be considered beneficial in the elderly as it could contribute to reinforce the intestinal barrier, to the increase in intestinal motility, to reduce inflammation and to increase the butyrate production by cross-feeding mechanisms with members of the intestinal microbiota able to produce this compound.

Relevant nutritional strategies for improving intestinal function include the design of diets adapted to the elderly as well as the use of selected probiotics, prebiotics, synbiotics (combined use of pro- and prebiotics), supplements and bioactive compounds. Probiotics are defined as 'live microorganisms, which when administered in adequate amounts confer health benefit to the host' (FAO/WHO, 2006). Most of the common probiotics are strains from different species of *Bifidobacterium* and *Lactobacillus*. Probiotics can act by inhibiting enteric pathogens or sequestrating their toxins (Valdés-Varela *et al.*, 2016), interacting with some members of the intestinal microbiota through cross-feeding mechanisms in the presence of prebiotic substrates (Rios-Covián *et al.*, 2015) and modulating the immune system directly or through their interaction with members of the gut microbiota (Hevia *et al.*, 2015). Prebiotics are 'selectively fermented ingredients resulting in specific changes in the composition and/or activity of the gastrointestinal microbiota, thus conferring benefits on host health' (Gibson *et al.*, 2017). Most prebiotics are complex carbohydrates or oligosaccharides from fruits, wholegrain s and other vegetables or are produced industrially. Prebiotics are not digested in the small intestine and they reach the colon unabsorbed, where they are selectively fermented by some members of the intestinal microbiota, resulting in beneficial changes in microbiota composition and its metabolic activity.

In spite of the increasing interest of the industry in the market of dietary products for the elderly, there is still a scarcity of studies demonstrating the efficacy of probiotics and prebiotics. Intervention studies carried out in the last years for demonstrating the efficacy of probiotic *Lactobacillus* and *Bifidobacterium* strains were mainly focused on ameliorating gastrointestinal disorders, such as constipation and diarrhoea associated with antibiotics or *C. difficile* infection, and few of them were carried out with the aim of improving the composition and functionality of the whole gut microbiota; nevertheless, some bifidobacteria and lactobacilli showed beneficial changes on specific groups of the intestinal microbiota and improved the profile of immune mediators (Bjorklund *et al.*, 2012; Salazar *et al.*, 2014, 2017; Spaiser *et al.*, 2015). Among prebiotics, galacto-oligosaccharides (GOS) were the most studied until now. A defined commercial mixture of trans-GOS, named as Bimuno or B-BOS, showed an improvement in the faecal microbiota composition and immune parameters in the elderly (Vulevic *et al.*, 2008, 2015). More human intervention studies are needed on the efficacy of alternative prebiotics such as xylo-oligosaccharides, beta-glucan, pectin, malto-oligosacharides and others. Synbiotic combinations of *Lactobacillus* and/or *Bifidobacterium* strains with prebiotic polysaccharides, oligosaccharides or polyols have been also tested to counteract health-related microbiota changes and impaired physiological functions in the elderly (Bartosch *et al.*, 2005; Bjorklund *et al.*, 2012; Macfarlane *et al.*, 2013). Polyphenols have been proven to promote beneficial changes in the intestinal microbiota and to exert health benefits on the host; however, the specific use in the elderly and the relationship between their metabolism by the intestinal microbiota and general host health parameters have been scarcely explored (Guadamuro *et al.*, 2017; Gutiérrez-Díaz *et al.*, 2017).

There is a need for more studies exploring the influence of diet and specific nutrients and supplements on microbiota functionality in the elderly. These studies should be performed with high number of individuals from different geographical locations in order to reach robust conclusions on the efficacy of nutritional interventions in the elderly.

Concluding remarks

The spectacular advance in recent years of DNA NGS technologies and its cost affordability for the analysis of complex microbial ecosystems has provided knowledge of the composition of the human intestinal microbiota, from its establishment in the early stages of life to the most advanced ages. Moreover, other omics technologies targeting mRNA, proteins and metabolites are allowing the metabolic potential of these microorganisms and their extraordinary influence on host health to be disentangled. The elderly population is increasing over the world and particularly in industrialized countries. The physiological, immunological and cognitive decline that occurs in old age is strongly influenced by lifestyle, social habits and nutritional status. Ageing is often accompanied by disease and disability. In addition, modification in eating patterns frequently occur, because of physiological and physical changes, which frequently involve a relative state of malnutrition that, in its turn, exerts a negative impact on the gut microbiota, the immune system and the cognitive state. All this results in a vicious circle that is difficult to leave. Current technological and scientific advances are bringing us the opportunity to understand the mechanisms underlying the beneficial or detrimental action of specific nutrients on the human intestinal microbiota and to know the influence of such microorganisms on physiological, immunological and cognitive function. It is therefore urgent that nutritional strategies are developed that are able to positively modulate the composition and functionality of the intestinal microbiota in order to promote healthy ageing that allow to dignify life in the elderly and to manage health-care costs.

References

AACC (2001). The definition of dietary fiber. Report of the Dietary Fiber Definition Committee to the Board of Directors of the American Association of Cereal Chemists *46*. Publication no. W-2001-0222-01. https://www.aaccnet.org/initiatives/definitions/Documents/DietaryFiber/DFDef.pdf.

Aagaard, K., Petrosino, J., Keitel, W., Watson, M., Katancik, J., Garcia, N., Patel, S., Cutting, M., Madden, T., Hamilton, H., *et al.* (2013). The Human Microbiome Project strategy for comprehensive sampling of the human microbiome and why it matters. FASEB J. *27*, 1012–1022. https://doi.org/10.1096/fj.12-220806.

Abram, F. (2015). Systems-based approaches to unravel multi-species microbial community functioning. Comput. Struct. Biotechnol. J. *13*, 24–32. https://doi.org/10.1016/j.csbj.2014.11.009.

Aliabadi, N., Lopman, B.A., Parashar, U.D., and Hall, A.J. (2015). Progress toward norovirus vaccines: Considerations for further development and implementation in potential target populations. Expert Rev. Vaccines *14*, 1241–1253. https://doi.org/10.1586/14760584.2015.1073110.

Amaral-Zettler, L.A., McCliment, E.A., Ducklow, H.W., and Huse, S.M. (2009). A method for studying protistan diversity using massively parallel sequencing of V9 hypervariable regions of small-subunit ribosomal RNA genes. PLOS ONE *4*, e6372. https://doi.org/10.1371/journal.pone.0006372.

Amarya, S., Singh, K., and Sabharwal, M. (2015). Changes during aging and their association with malnutrition. J. Clin. Gerontol. Geriatr. *6*, 78–84. https://dx.doi.org/10.1016/j.jcgg.2015.05.003.

Arboleya, S., Binetti, A., Salazar, N., Fernández, N., Solís, G., Hernández-Barranco, A., Margolles, A., de Los Reyes-Gavilán, C.G., and Gueimonde, M. (2012). Establishment and development of intestinal microbiota in preterm neonates. FEMS Microbiol. Ecol. *79*, 763–772. https://doi.org/10.1111/j.1574-6941.2011.01261.x.

Arumugam, M., Raes, J., Pelletier, E., Le Paslier, D., Yamada, T., Mende, D.R., Fernandes, G.R., Tap, J., Bruls, T., Batto, J.M., *et al.* (2011). Enterotypes of the human gut microbiome. Nature *473*, 174–180. https://doi.org/10.1038/nature09944.

Asempa, T.E., and Nicolau, D.P. (2017). *Clostridium difficile* infection in the elderly: An update on management. Clin. Interv. Aging *12*, 1799–1809. https://doi.org/10.2147/CIA.S149089.

Atarashi, K., Tanoue, T., Oshima, K., Suda, W., Nagano, Y., Nishikawa, H., Fukuda, S., Saito, T., Narushima, S., Hase, K., *et al.* (2013). Treg induction by a rationally selected mixture of Clostridia strains from the human microbiota. Nature *500*, 232–236. https://doi.org/10.1038/nature12331.

Bagdasarian, N., Rao, K., and Malani, P.N. (2015). Diagnosis and treatment of *Clostridium difficile* in adults: A systematic review. JAMA *313*, 398–408. https://doi.org/10.1001/jama.2014.17103.

Bartosch, S., Fite, A., Macfarlane, G.T., and McMurdo, M.E. (2004). Characterization of bacterial communities in feces from healthy elderly volunteers and hospitalized elderly patients by using real-time PCR and effects of antibiotic treatment on the fecal microbiota. Appl. Environ. Microbiol. *70*, 3575–3581. https://doi.org/10.1128/AEM.70.6.3575-3581.2004.

Bartosch, S., Woodmansey, E.J., Paterson, J.C., McMurdo, M.E., and Macfarlane, G.T. (2005). Microbiological effects of consuming a synbiotic containing *Bifidobacterium bifidum, Bifidobacterium lactis,* and oligofructose in elderly persons, determined by real-time polymerase chain reaction and counting of viable bacteria. Clin. Infect. Dis. *40,* 28–37. http://doi.org/10.1086/426027.

Beaufrère, B., and Morio, B. (2000). Fat and protein redistribution with aging: Metabolic considerations. Eur. J. Clin. Nutr. *54* (Suppl. 3), S48–53.

Bernstein, M.A., Tucker, K.L., Ryan, N.D., O'Neill, E.F., Clements, K.M., Nelson, M.E., Evans, W.J., and Fiatarone Singh, M.A. (2002). Higher dietary variety is associated with better nutritional status in frail elderly people. J. Am. Diet. Assoc. *102,* 1096–1104. https://doi.org/10.1016/S0002-8223(02)80078-5.

Biagi, E., Candela, M., Fairweather-Tait, S., Franceschi, C., and Brigidi, P. (2012). Aging of the human metaorganism: The microbial counterpart. Age *34,* 247–267. https://doi.org/10.1007/s11357-011-9217-5.

Biagi, E., Franceschi, C., Rampelli, S., Severgnini, M., Ostan, R., Turroni, S., Consolandi, C., Quercia, S., Scurti, M., Monti, D., *et al.* (2016). Gut microbiota and extreme longevity. Curr. Biol. *26,* 1480–1485. https://doi.org/10.1016/j.cub.2016.04.016.

Biagi, E., Nylund, L., Candela, M., Ostan, R., Bucci, L., Pini, E., Nikkïla, J., Monti, D., Satokari, R., Franceschi, C., *et al.* (2010). Through ageing, and beyond: Gut microbiota and inflammatory status in seniors and centenarians. PLOS ONE *5,* e10667. https://doi.org/10.1371/journal.pone.0010667.

Björklund, M., Ouwehand, A.C., Forssten, S.D., Nikkilä, J., Tiihonen, K., Rautonen, N., and Lahtinen, S.J. (2012). Gut microbiota of healthy elderly NSAID users is selectively modified with the administration of *Lactobacillus acidophilus* NCFM and lactitol. Age *34,* 987–999. https://doi.org/10.1007/s11357-011-9294-5.

Calvani, R., Brasili, E., Praticò, G., Capuani, G., Tomassini, A., Marini, F., Sciubba, F., Finamore, A., Roselli, M., Marzetti, E., *et al.* (2014). Fecal and urinary NMR-based metabolomics unveil an aging signature in mice. Exp. Gerontol. *49,* 5–11. https://doi.org/10.1016/j.exger.2013.10.010.

Camp, J.G., Frank, C.L., Lickwar, C.R., Guturu, H., Rube, T., Wenger, A.M., Chen, J., Bejerano, G., Crawford, G.E., and Rawls, J.F. (2014). Microbiota modulate transcription in the intestinal epithelium without remodeling the accessible chromatin landscape. Genome Res. *24,* 1504–1516. https://doi.org/10.1101/gr.165845.113.

Candore, G., Balistreri, C.R., Colonna-Romano, G., Grimaldi, M.P., Lio, D., Listi', F., Scola, L., Vasto, S., and Caruso, C. (2008). Immunosenescence and anti-immunosenescence therapies: The case of probiotics. Rejuvenation Res. *11,* 425–432. https://doi.org/10.1089/rej.2008.0662.

Candore, G., Caruso, C., Jirillo, E., Magrone, T., and Vasto, S. (2010). Low grade inflammation as a common pathogenic denominator in age-related diseases: Novel drug targets for anti-ageing strategies and successfully ageing achievement. Curr. Pharm. Des. *16,* 584–596. https://dx.doi.org/10.2174/138161210790883868.

Castellarin, M., Warren, R.L., Freeman, J.D., Dreolini, L., Krzywinski, M., Strauss, J., Barnes, R., Watson, P., Allen-Vercoe, E., Moore, R.A., *et al.* (2012). *Fusobacterium nucleatum* infection is prevalent in human colorectal carcinoma. Genome Res. *22,* 299–306. https://doi.org/10.1101/gr.126516.111.

Chun, O.K., Chung, S.J., and Song, W.O. (2007). Estimated dietary flavonoid intake and major food sources of U.S. adults. J. Nutr. *137,* 1244–1252.

Claesson, M.J., Cusack, S., O'Sullivan, O., Greene-Diniz, R., de Weerd, H., Flannery, E., Marchesi, J.R., Falush, D., Dinan, T., Fitzgerald, G., *et al.* (2011). Composition, variability, and temporal stability of the intestinal microbiota of the elderly. Proc. Natl. Acad. Sci. U.S.A. *108* (Suppl. 1), 4586–4591. https://doi.org/10.1073/pnas.1000097107.

Claesson, M.J., Jeffery, I.B., Conde, S., Power, S.E., O'Connor, E.M., Cusack, S., Harris, H.M., Coakley, M., Lakshminarayanan, B., O'Sullivan, O., *et al.* (2012). Gut microbiota composition correlates with diet and health in the elderly. Nature *488,* 178–184. https://doi.org/10.1038/nature11319.

Clemente, J.C., Ursell, L.K., Parfrey, L.W., and Knight, R. (2012). The impact of the gut microbiota on human health: An integrative view. Cell *148,* 1258–1270. https://doi.org/10.1016/j.cell.2012.01.035.

Collado, M.C., Rautava, S., Aakko, J., Isolauri, E., and Salminen, S. (2016). Human gut colonisation may be initiated *in utero* by distinct microbial communities in the placenta and amniotic fluid. Sci. Rep. *6,* 23129. https://doi.org/10.1038/srep23129.

Collino, S., Montoliu, I., Martin, F.P., Scherer, M., Mari, D., Salvioli, S., Bucci, L., Ostan, R., Monti, D., Biagi, E., *et al.* (2013). Metabolic signatures of extreme longevity in northern Italian centenarians reveal a complex remodeling of lipids, amino acids, and gut microbiota metabolism. PLOS ONE *8,* e56564. https://doi.org/10.1371/journal.pone.0056564.

Cuervo, A., Valdés, L., Salazar, N., de los Reyes-Gavilán, C.G., Ruas-Madiedo, P., Gueimonde, M., and González, S. (2014). Pilot study of diet and microbiota: Interactive associations of fibers and polyphenols with human intestinal bacteria. J. Agric. Food Chem. *62,* 5330–5336. https://doi.org/10.1021/jf501546a.

Culross, B. (2008). Gerontology update. Nutrition: Meeting the needs of the elderly. ARN Network August/September, 7. Available online: https://es.scribd.com/document/222822480/Meeting-the-Nutrition-Needs-of-Elderly (accessed 25 September 2018).

De Filippis, F., Pellegrini, N., Vannini, L., Jeffery, I.B., La Storia, A., Laghi, L., Serrazanetti, D.I., Di Cagno, R., Ferrocino, I., Lazzi, C., *et al.* (2016). High-level adherence to a Mediterranean diet beneficially impacts the gut microbiota and associated metabolome. Gut *65,* 1812–1821. https://doi.org/10.1136/gutjnl-2015-309957.

Dolgin, E. (2016). How to defeat dementia. Nature *539,* 156–158. https://doi.org/10.1038/539156a.

Duytschaever, G., Huys, G., Bekaert, M., Boulanger, L., De Boeck, K., and Vandamme, P. (2013). Dysbiosis of bifidobacteria and *Clostridium* cluster XIVa in the cystic fibrosis fecal microbiota. J. Cyst. Fibros. *12,* 206–215. https://doi.org/10.1016/j.jcf.2012.10.003.

Eckburg, P.B., Bik, E.M., Bernstein, C.N., Purdom, E., Dethlefsen, L., Sargent, M., Gill, S.R., Nelson, K.E., and Relman, D.A. (2005). Diversity of the human intestinal microbial flora. Science *308,* 1635–1638.

Elliott, B., Androga, G.O., Knight, D.R., and Riley, T.V. (2017). *Clostridium difficile* infection: Evolution, phylogeny and molecular epidemiology. Infect. Genet. Evol. *49,* 1–11.

Escherich, T. (1989). The intestinal bacteria of the neonate and breast-fed infant. 1885. Rev. Infect. Dis. *11,* 352–356. https://doi.org/10.1093/clinids/11.2.352.

Fanaro, S., Chierici, R., Guerrini, P., and Vigi, V. (2003). Intestinal microflora in early infancy: Composition and development. Acta Paediatr. Suppl. *91,* 48–55.

FAO/ WHO. (2006). Probiotics in food: Health and nutritional properties and guidelines for evaluation. FAO Food and Nutrition Paper *85.* Available online: http://www.fao.org/3/a-a0512e.pdf. Accessed 16 August 2018.

Fiehn, O. (2002). Metabolomics – the link between genotypes and phenotypes. Plant Mol. Biol. *48,* 155–171.

Flint, H.J. (2012). The impact of nutrition on the human microbiome. Nutr. Rev. *70* (Suppl. 1), S10–3. https://doi.org/10.1111/j.1753-4887.2012.00499.x.

Franceschi, C., Bonafè, M., Valensin, S., Olivieri, F., De Luca, M., Ottaviani, E., and De Benedictis, G. (2000). Inflamm-aging. An evolutionary perspective on immunosenescence. Ann. N. Y. Acad. Sci. *908,* 244–254.

Franzosa, E.A., Morgan, X.C., Segata, N., Waldron, L., Reyes, J., Earl, A.M., Giannoukos, G., Boylan, M.R., Ciulla, D., Gevers, D., *et al.* (2014). Relating the metatranscriptome and metagenome of the human gut. Proc. Natl. Acad. Sci. U.S.A. *111,* E2329–38. https://doi.org/10.1073/pnas.1319284111.

Fukuda, S., Toh, H., Hase, K., Oshima, K., Nakanishi, Y., Yoshimura, K., Tobe, T., Clarke, J.M., Topping, D.L., Suzuki, T., *et al.* (2011). Bifidobacteria can protect from enteropathogenic infection through production of acetate. Nature *469,* 543–547. https://doi.org/10.1038/nature09646.

Gao, X.Y., Kuang, H.Y., Liu, X.M., Duan, P., Yang, Y., and Ma, Z.B. (2009a). Circulating ghrelin/obestatin ratio in subjects with *Helicobacter pylori* infection. Nutrition *25,* 506–511. https://doi.org/10.1016/j.nut.2008.11.002.

Gao, Z., Yin, J., Zhang, J., Ward, R.E., Martin, R.J., Lefevre, M., Cefalu, W.T., and Ye, J. (2009b). Butyrate improves insulin sensitivity and increases energy expenditure in mice. Diabetes *58,* 1509–1517. https://doi.org/10.2337/db08-1637.

Garrett, W.S., Gordon, J.I., and Glimcher, L.H. (2010). Homeostasis and inflammation in the intestine. Cell *140,* 859–870. https://doi.org/10.1016/j.cell.2010.01.023.

Gavini, F., Cayuela, Ch., Antoine, J.M., Lecoq, C., Lefebvre, B., Membré, J.M., and Neut, C. (2001). Differences in the distribution of bifidobacterial and enterobacterial species in human faecal microflora of three different (children, adults, elderly) age groups. Microb. Ecol. Health Dis. *13,* 40–45. https://dx.doi.org/10.1080/089106001750071690.

Genta, R.M., Turner, K.O., and Sonnenberg, A. (2017). Demographic and socioeconomic influences on *Helicobacter pylori* gastritis and its pre-neoplastic lesions amongst US residents. Aliment. Pharmacol. Ther. *46,* 322–330. https://doi.org/10.1111/apt.14162.

Gibson, G.R., Hutkins, R., Sanders, M.E., Prescott, S.L., Reimer, R.A., Salminen, S.J., Scott, K., Stanton, C., Swanson, K.S., Cani, P.D., *et al.* (2017). Expert consensus document: The International Scientific Association for Probiotics and Prebiotics (ISAPP) consensus statement on the definition and scope of prebiotics. Nat. Rev. Gastroenterol. Hepatol. *14,* 491–502. https://doi.org/10.1038/nrgastro.2017.75.

Gill, S.R., Pop, M., Deboy, R.T., Eckburg, P.B., Turnbaugh, P.J., Samuel, B.S., Gordon, J.I., Relman, D.A., Fraser-Liggett, C.M., and Nelson, K.E. (2006). Metagenomic analysis of the human distal gut microbiome. Science *312*, 1355–1359.

Gonsalves, W.C., Wrightson, A.S., and Henry, R.G. (2008). Common oral conditions in older persons. Am. Fam. Physician *78*, 845–852.

Gosalbes, M.J., Durbán, A., Pignatelli, M., Abellan, J.J., Jiménez-Hernández, N., Pérez-Cobas, A.E., Latorre, A., and Moya, A. (2011). Metatranscriptomic approach to analyze the functional human gut microbiota. PLOS ONE *6*, e17447. https://doi.org/10.1371/journal.pone.0017447.

Goulet, O. (2015). Potential role of the intestinal microbiota in programming health and disease. Nutr. Rev. *73* (Suppl. 1), 32–40. https://doi.org/10.1093/nutrit/nuv039.

Guadamuro, L., Dohrmann, A.B., Tebbe, C.C., Mayo, B., and Delgado, S. (2017). Bacterial communities and metabolic activity of faecal cultures from equol producer and non-producer menopausal women under treatment with soy isoflavones. BMC Microbiol. *17*, 93. https://doi.org/10.1186/s12866-017-1001-y.

Guglielmetti, S., Fracassetti, D., Taverniti, V., Del Bo', C., Vendrame, S., Klimis-Zacas, D., Arioli, S., Riso, P., and Porrini, M. (2013). Differential modulation of human intestinal bifidobacterium populations after consumption of a wild blueberry (*Vaccinium angustifolium*) drink. J. Agric. Food Chem. *61*, 8134–8140. https://doi.org/10.1021/jf402495k.

Gutiérrez-Díaz, I., Fernández-Navarro, T., Salazar, N., Bartolomé, B., Moreno-Arribas, M.V., de Andres-Galiana, E.J., Fernández-Martínez, J.L., de Los Reyes-Gavilán, C.G., Gueimonde, M., and González, S. (2017). Adherence to a Mediterranean diet influences the fecal metabolic profile of microbial-derived phenolics in a Spanish cohort of middle-age and older people. J. Agric. Food Chem. *65*, 586–595. https://doi.org/10.1021/acs.jafc.6b04408.

Hall, A.J., Curns, A.T., McDonald, L.C., Parashar, U.D., and Lopman, B.A. (2012). The roles of *Clostridium difficile* and norovirus among gastroenteritis-associated deaths in the United States, 1999–2007. Clin. Infect. Dis. *55*, 216–223. https://doi.org/10.1093/cid/cis386.

Hayashi, H., Sakamoto, M., Kitahara, M., and Benno, Y. (2003). Molecular analysis of fecal microbiota in elderly individuals using 16S rDNA library and T-RFLP. Microbiol. Immunol. *47*, 557–570.

Hevia, A., Delgado, S., Sánchez, B., and Margolles, A. (2015). Molecular players involved in the interaction between beneficial bacteria and the immune system. Front. Microbiol. *6*, 1285. https://doi.org/10.3389/fmicb.2015.01285.

Hooper, L.V., and Gordon, J.I. (2001). Commensal host-bacterial relationships in the gut. Science *292*, 1115–1118.

Hopkins, M.J., Sharp, R., and Macfarlane, G.T. (2001). Age and disease related changes in intestinal bacterial populations assessed by cell culture, 16S rRNA abundance, and community cellular fatty acid profiles. Gut *48*, 198–205.

Human Microbiome Project Consortium (2012). Structure, function and diversity of the healthy human microbiome. Nature *486*, 207–214. https://dx.doi.org/ 10.1038/nature11234.

Hungate, R.E. (1969). A roll tube method for cultivation of strict anaerobes. In Methods in Microbiology, Norris, J.R, and Ribbons, D.W., eds. (Academic Press, London), vol. 3b, pp. 117–132.

Huse, S.M., Ye, Y., Zhou, Y., and Fodor, A.A. (2012). A core human microbiome as viewed through 16S rRNA sequence clusters. PLOS ONE 7, e34242. https://doi.org/10.1371/journal.pone.0034242.

IOM (Institute of Medicine). (2010). Providing Healthy and Safe Foods as We Age: Workshop Summary (The National Academies Press, Washington, DC).

Jackson, M.A., Jackson, M., Jeffery, I.B., Beaumont, M., Bell, J.T., Clark, A.G., Ley, R.E., O'Toole, P.W., Spector, T.D., and Steves, C.J. (2016). Signatures of early frailty in the gut microbiota. Genome Med. *8*, 8. https://doi.org/10.1186/s13073-016-0262-7.

Jeffery, I.B., Lynch, D.B., and O'Toole, P.W. (2016). Composition and temporal stability of the gut microbiota in older persons. ISME J. *10*, 170–182. https://doi.org/10.1038/ismej.2015.88.

Jing, Y., Gravenstein, S., Chaganty, N.R., Chen, N., Lyerly, K.H., Joyce, S., and Deng, Y. (2007). Aging is associated with a rapid decline in frequency, alterations in subset composition, and enhanced Th2 response in CD1d-restricted NKT cells from human peripheral blood. Exp. Gerontol. *42*, 719–732.

Kasai, C., Sugimoto, K., Moritani, I., Tanaka, J., Oya, Y., Inoue, H., Tameda, M., Shiraki, K., Ito, M., Takei, Y., *et al.* (2015). Comparison of the gut microbiota composition between obese and non-obese individuals in a Japanese population, as analyzed by terminal restriction fragment length polymorphism and next-generation sequencing. BMC Gastroenterol. *15*, 100. https://doi.org/10.1186/s12876-015-0330-2.

Kennedy, B.K., Berger, S.L., Brunet, A., Campisi, J., Cuervo, A.M., Epel, E.S., Franceschi, C., Lithgow, G.J., Morimoto, R.I., Pessin, J.E., *et al.* (2014). Geroscience: Linking aging to chronic disease. Cell *159*, 709–713. https://doi.org/10.1016/j.cell.2014.10.039.

Kim, B.J., Kim, H.Y., Yun, Y.J., Kim, B.J., and Kook, Y.H. (2010). Differentiation of *Bifidobacterium* species using partial RNA polymerase β-subunit (*rpoB*) gene sequences. Int. J. Syst. Evol. Microbiol. *60*, 2697–2704. https://dx.doi.org/ 10.1099/ijs.0.020339-0.

Kim, S.H., Kim, J.W., Byun, J., Jeong, J.B., Kim, B.G., and Lee, K.L. (2017). Plasma ghrelin level and plasma ghrelin/obestatin ratio are related to intestinal metaplasia in elderly patients with functional dyspepsia. PLOS ONE *12*, e0175231. https://doi.org/10.1371/journal.pone.0175231.

Kitahara, M., Takamine, F., Imamura, T., and Benno, Y. (2000). Assignment of *Eubacterium* sp. VPI 12708 and related strains with high bile acid 7 alpha-dehydroxylating activity to *Clostridium scindens* and proposal of *Clostridium hylemonae* sp nov., isolated from human faeces. Int. J. Syst. Evol. Microbiol. *50*, 971–978. https://dx.doi.org/10.1099/00207713-50-3-97.

Kociolek, L.K., and Gerding, D.N. (2016). Breakthroughs in the treatment and prevention of *Clostridium difficile* infection. Nat. Rev. Gastroenterol. Hepatol. *13*, 150–160. https://doi.org/10.1038/nrgastro.2015.220.

Kullin, B., Meggersee, R., D'Alton, J., Galvao, B., Rajabally, N., Whitelaw, A., Bamford, C., Reid, S.J., and Abratt, V.R. (2015). Prevalence of gastrointestinal pathogenic bacteria in patients with diarrhoea attending Groote Schuur Hospital, Cape Town, South Africa. S. Afr. Med. J. *105*, 121–125. https://doi.org/10.7196/samj.8654.

Kumar, M., Babaei, P., Ji, B., and Nielsen, J. (2016). Human gut microbiota and healthy aging: Recent developments and future prospective. Nutr. Healthy Aging *4*, 3–16. https://doi.org/10.3233/NHA-150002.

Lagier, J.C., Armougom, F., Million, M., Hugon, P., Pagnier, I., Robert, C., Bittar, F., Fournous, G., Gimenez, G., Maraninchi, M., *et al.* (2012). Microbial culturomics: paradigm shift in the human gut microbiome study. Clin. Microbiol. Infect. *18*, 1185–1193. https://doi.org/10.1111/1469-0691.12023.

Lagier, J.C., Edouard, S., Pagnier, I., Mediannikov, O., Drancourt, M., and Raoult, D. (2015). Current and past strategies for bacterial culture in clinical microbiology. Clin. Microbiol. Rev. *28*, 208–236. https://doi.org/10.1128/CMR.00110-14.

Lagier, J.C., Khelaifia, S., Alou, M.T., Ndongo, S., Dione, N., Hugon, P., Caputo, A., Cadoret, F., Traore, S.I., Seck, E.H., *et al.* (2016). Culture of previously uncultured members of the human gut microbiota by culturomics. Nat. Microbiol. *1*, 16203. https://doi.org/10.1038/nmicrobiol.2016.203.

La-Ongkham, O., Nakphaichit, M., Leelavatcharamas, V., Keawsompong, S., and Nitisinprasert, S. (2015). Distinct gut microbiota of healthy children from two different geographic regions of Thailand. Arch. Microbiol. *197*, 561–573. https://doi.org/10.1007/s00203-015-1089-0.

Larbi, A., Franceschi, C., Mazzatti, D., Solana, R., Wikby, A., and Pawelec, G. (2008). Aging of the immune system as a prognostic factor for human longevity. Physiology *23*, 64–74. https://doi.org/10.1152/physiol.00040.2007.

Layé, S., Delpech, J-C., De Smedt-Peyrusse, V., Joffre, C., Larrieu, T., Madore, C., Nadjar, A., and Capuron, L. (2011). Neuroinflammation and aging: influence of dietary n-3 polyunsaturated fatty acid. OCL *18*, 301–306. https://dx.doi.org/10.1051/ocl.2011.0412.

Lee, H.C., Jenner, A.M., Low, C.S., and Lee, Y.K. (2006). Effect of tea phenolics and their aromatic fecal bacterial metabolites on intestinal microbiota. Res. Microbiol. *157*, 876–884.

Lee, P.Y., Chin, S.F., Neoh, H.M., and Jamal, R. (2017). Metaproteomic analysis of human gut microbiota: Where are we heading? J. Biomed. Sci. *24*, 36. https://doi.org/10.1186/s12929-017-0342-z.

Lewis, B.B., Buffie, C.G., Carter, R.A., Leiner, I., Toussaint, N.C., Miller, L.C., Gobourne, A., Ling, L., and Pamer, E.G. (2015). Loss of microbiota-mediated colonization resistance to *Clostridium difficile* infection with oral vancomycin compared with metronidazole. J. Infect. Dis. *212*, 1656–1665. https://doi.org/10.1093/infdis/jiv256.

Li, D., Wang, P., Wang, P., Hu, X., and Chen, F. (2016). The gut microbiota: A treasure for human health. Biotechnol. Adv. *34*, 1210–1224.

Li, D., Wang, P., Wang, P., Hu, X., and Chen, F. (2017). Targeting the gut microbiota by dietary nutrients: A new avenue for human health. Crit. Rev. Food Sci. Nutr. [Epub ahead of print]. https://doi.org/10.1080/10408398.2017.1363708.

Li, J., Jia, H., Cai, X., Zhong, H., Feng, Q., Sunagawa, S., Arumugam, M., Kultima, J.R., Prifti, E., Nielsen, T., *et al.* (2014). An integrated catalog of reference genes in the human gut microbiome. Nat. Biotechnol. *32*, 834–841. https://doi.org/10.1038/nbt.2942.

Lloyd-Price, J., Mahurkar, A., Rahnavard, G., Crabtree, J., Orvis, J., Hall, A.B., Brady, A., Creasy, H.H., McCracken, C., Giglio, M.G., *et al.* (2017). Strains, functions and dynamics in the expanded Human Microbiome Project. Nature *550*, 61–66. https://doi.org/10.1038/nature23889.

Louis, P., and Flint, H.J. (2009). Diversity, metabolism and microbial ecology of butyrate-producing bacteria from the human large intestine. FEMS Microbiol. Lett. *294*, 1–8. https://doi.org/10.1111/j.1574-6968.2009.01514.x.

Louis, P., Young, P., Holtrop, G., and Flint, H.J. (2010). Diversity of human colonic butyrate-producing bacteria revealed by analysis of the butyryl-CoA:acetate CoA-transferase gene. Environ. Microbiol. *12*, 304–314. https://doi.org/10.1111/j.1462-2920.2009.02066.x.

Lozupone, C.A., Stombaugh, J.I., Gordon, J.I., Jansson, J.K., and Knight, R. (2012). Diversity, stability and resilience of the human gut microbiota. Nature *489*, 220–230. https://doi.org/10.1038/nature11550.

Luo, G., Zhang, Y., Guo, P., Wang, L., Huang, Y., and Li, K. (2017). Global patterns and trends in stomach cancer incidence: Age, period and birth cohort analysis. Int. J. Cancer *141*, 1333–1344. https://doi.org/10.1002/ijc.30835.

Lynch, D.B., Jeffery, I.B., Cusack, S., O'Connor, E.M., and O'Toole, P.W. (2014). Diet-microbiota-health interactions in older subjects: Implications for healthy aging. In Aging and Health – A Systems Biology Perspective, Yashin, A.I., and Jazwinski, S.M., eds. (Karger AG, Basel), vol. 40, pp. 141–154. https://dx.doi.org/10.1159/000364976.

Macfarlane, S., Cleary, S., Bahrami, B., Reynolds, N., and Macfarlane, G.T. (2013). Synbiotic consumption changes the metabolism and composition of the gut microbiota in older people and modifies inflammatory processes: A randomised, double-blind, placebo-controlled crossover study. Aliment. Pharmacol. Ther. *38*, 804–816. https://doi.org/10.1111/apt.12453.

Magill, S.S., Edwards, J.R., Bamberg, W., Beldavs, Z.G., Dumyati, G., Kainer, M.A., Lynfield, R., Maloney, M., McAllister-Hollod, L., Nadle, J., *et al.* (2014). Multistate point-prevalence survey of health care-associated infections. N. Engl. J. Med. *370*, 1198–1208. https://doi.org/10.1056/NEJMoa1306801.

Mäkivuokko, H., Tiihonen, K., Tynkkynen, S., Paulin, L., and Rautonen, N. (2010). The effect of age and non-steroidal anti-inflammatory drugs on human intestinal microbiota composition. Br. J. Nutr. *103*, 227–234. https://doi.org/10.1017/S0007114509991553.

Malfertheiner, P., Megraud, F., O'Morain, C.A., Gisbert, J.P., Kuipers, E.J., Axon, A.T., Bazzoli, F., Gasbarrini, A., Atherton, J., Graham, D.Y., *et al.* (2017). Management of *Helicobacter pylori* infection-the Maastricht V/Florence Consensus Report. Gut *66*, 6–30. https://doi.org/10.1136/gutjnl-2016-312288.

Martinez-Medina, M., Aldeguer, X., Gonzalez-Huix, F., Acero, D., and Garcia-Gil, L.J. (2006). Abnormal microbiota composition in the ileocolonic mucosa of Crohn's disease patients as revealed by polymerase chain reaction-denaturing gradient gel electrophoresis. Inflamm. Bowel Dis. *12*, 1136–1145. https://dx.doi.org/ 10.1097/01.mib.0000235828.09305.0c.

Maslowski, K.M., and Mackay, C.R. (2011). Diet, gut microbiota and immune responses. Nat. Immunol. *12*, 5–9. https://doi.org/10.1038/ni0111-5.

Massot-Cladera, M., Pérez-Berezo, T., Franch, A., Castell, M., and Pérez-Cano, F.J. (2012). Cocoa modulatory effect on rat faecal microbiota and colonic crosstalk. Arch. Biochem. Biophys. *527*, 105–112. https://doi.org/10.1016/j.abb.2012.05.015.

Mazari, L., and Lesourd, B.M. (1998). Nutritional influences on immune response in healthy aged persons. Mech. Ageing Dev. *104*, 25–40.

McCarty, M.F., and DiNicolantonio, J.J. (2015). An increased need for dietary cysteine in support of glutathione synthesis may underlie the increased risk for mortality associated with low protein intake in the elderly. Age *37*, 96. https://doi.org/10.1007/s11357-015-9823-8.

Milani, C., Ticinesi, A., Gerritsen, J., Nouvenne, A., Lugli, G.A., Mancabelli, L., Turroni, F., Duranti, S., Mangifesta, M., Viappiani, A., *et al.* (2016). Gut microbiota composition and *Clostridium difficile* infection in hospitalized elderly individuals: A metagenomic study. Sci. Rep. *6*, 25945. https://doi.org/10.1038/srep25945.

Miller, C.P., Bohnhoff, M., and Rifkind, D. (1957). The effect of an antibiotic on the susceptibility of the mouse's intestinal tract to *Salmonella* infection. Trans. Am. Clin. Climatol. Assoc. *68*, 51–58.

Monti, D., Ostan, R., Borelli, V., Castellani, G., and Franceschi, C. (2017). Inflammaging and human longevity in the omics era. Mech. Ageing Dev. *165*, 129–138. https://dx.doi.org/10.1016/j.mad.2016.12.008.

Morgan, X.C., Tickle, T.L., Sokol, H., Gevers, D., Devaney, K.L., Ward, D.V., Reyes, J.A., Shah, S.A., LeLeiko, N., Snapper, S.B., *et al.* (2012). Dysfunction of the intestinal microbiome in inflammatory bowel disease and treatment. Genome Biol. *13*, R79. https://doi.org/10.1186/gb-2012-13-9-r79.

Muegge, B.D., Kuczynski, J., Knights, D., Clemente, J.C., González, A., Fontana, L., Henrissat, B., Knight, R., and Gordon, J.I. (2011). Diet drives convergence in gut microbiome functions across mammalian phylogeny and within humans. Science *332*, 970–974. https://doi.org/10.1126/science.1198719.

Mueller, S., Saunier, K., Hanisch, C., Norin, E., Alm, L., Midtvedt, T., Cresci, A., Silvi, S., Orpianesi, C., Verdenelli, M.C., *et al.* (2006). Differences in fecal microbiota in different European study populations in relation to age, gender, and country: A cross-sectional study. Appl. Environ. Microbiol. *72*, 1027–1033.

Murphy, K., O' Shea, C.A., Ryan, C.A., Dempsey, E.M., O' Toole, P.W., Stanton, C., and Ross, R.P. (2015). The gut microbiota composition in dichorionic triplet sets suggests a role for host genetic factors. PLOS ONE *10*, e0122561. https://doi.org/10.1371/journal.pone.0122561.

Nogacka, A., Salazar, N., Suárez, M., Milani, C., Arboleya, S., Solís, G., Fernández, N., Alaez, L., Hernández-Barranco, A.M., de Los Reyes-Gavilán, C.G., *et al.* (2017). Impact of intrapartum antimicrobial prophylaxis upon the intestinal microbiota and the prevalence of antibiotic resistance genes in vaginally delivered full-term neonates. Microbiome *5*, 93. https://doi.org/10.1186/s40168-017-0313-3.

Nogacka, A.M., Salazar, N., Arboleya, S., Suárez, M., Fernández, N., Solís, G., de Los Reyes-Gavilán, C.G., and Gueimonde, M. (2018). Early microbiota, antibiotics and health. Cell. Mol. Life Sci. *75*, 83–91. https://doi.org/10.1007/s00018-017-2670-2.

O'Toole, P.W., and Claesson, M.J. (2010). Gut microbiota : Changes throughout the lifespan from infancy to elderly. Int. Dairy J. *20*, 281–291. https://dx.doi.org/10.1016/j.idairyj.2009.11.010.

O'Toole, P.W., and Jeffery, I.B. (2018). Microbiome-health interactions in older people. Cell. Mol. Life Sci. *75*, 119–128. https://doi.org/10.1007/s00018-017-2673-z.

Odamaki, T., Kato, K., Sugahara, H., Hashikura, N., Takahashi, S., Xiao, J.Z., Abe, F., and Osawa, R. (2016). Age-related changes in gut microbiota composition from newborn to centenarian: A cross-sectional study. BMC Microbiol. *16*, 90. https://doi.org/10.1186/s12866-016-0708-5.

Oulas, A., Pavloudi, C., Polymenakou, P., Pavlopoulos, G.A., Papanikolaou, N., Kotoulas, G., Arvanitidis, C., and Iliopoulos, I. (2015). Metagenomics: Tools and insights for analyzing next-generation sequencing data derived from biodiversity studies. Bioinform. Biol. Insights *9*, 75–88. https://doi.org/10.4137/BBI.S12462.

Peng, L., Li, Z.R., Green, R.S., Holzman, I.R., and Lin, J. (2009). Butyrate enhances the intestinal barrier by facilitating tight junction assembly via activation of AMP-activated protein kinase in Caco-2 cell monolayers. J. Nutr. *139*, 1619–1625. https://doi.org/10.3945/jn.109.104638.

Petry, N., Egli, I., Chassard, C., Lacroix, C., and Hurrell, R. (2012). Inulin modifies the bifidobacteria population, fecal lactate concentration, and fecal pH but does not influence iron absorption in women with low iron status. Am. J. Clin. Nutr. *96*, 325–331. https://dx.doi.org/10.3945/ajcn.112.035717.

Prag, C., Prag, M., and Fredlund, H. (2017). Proton pump inhibitors as a risk factor for norovirus infection. Epidemiol. Infect. *145*, 1617–1623. https://doi.org/10.1017/S0950268817000528.

Prakash, T., and Taylor, T.D. (2012). Functional assignment of metagenomic data: Challenges and applications. Brief. Bioinformatics *13*, 711–727. https://doi.org/10.1093/bib/bbs033.

Qin, J., Li, R., Raes, J., Arumugam, M., Burgdorf, K.S., Manichanh, C., Nielsen, T., Pons, N., Levenez, F., Yamada, T., *et al.* (2010). A human gut microbial gene catalogue established by metagenomic sequencing. Nature *464*, 59–65. https://doi.org/10.1038/nature08821.

Queipo-Ortuño, M.I., Boto-Ordóñez, M., Murri, M., Gomez-Zumaquero, J.M., Clemente-Postigo, M., Estruch, R., Cardona Diaz, F., Andrés-Lacueva, C., and Tinahones, F.J. (2012). Influence of red wine polyphenols and ethanol on the gut microbiota ecology and biochemical biomarkers. Am. J. Clin. Nutr. *95*, 1323–1334. https://doi.org/10.3945/ajcn.111.027847.

Rajilić-Stojanović, M., and de Vos, W.M. (2014). The first 1000 cultured species of the human gastrointestinal microbiota. FEMS Microbiol. Rev. *38*, 996–1047. https://doi.org/10.1111/1574-6976.12075.

Rampelli, S., Candela, M., Turroni, S., Biagi, E., Collino, S., Franceschi, C., O'Toole, P.W., and Brigidi, P. (2013). Functional metagenomic profiling of intestinal microbiome in extreme ageing. Aging *5*, 902–912.

Rand, K.H., Tremblay, E.E., Hoidal, M., Fisher, L.B., Grau, K.R., and Karst, S.M. (2015). Multiplex gastrointestinal pathogen panels: Implications for infection control. Diagn. Microbiol. Infect. Dis. *82*, 154–157. https://doi.org/10.1016/j.diagmicrobio.2015.01.007.

Rea, M.C., O'Sullivan, O., Shanahan, F., O'Toole, P.W., Stanton, C., Ross, R.P., and Hill, C. (2012). *Clostridium difficile* carriage in elderly subjects and associated changes in the intestinal microbiota. J. Clin. Microbiol. *50*, 867–875. https://doi.org/10.1128/JCM.05176-11.

Reichardt, N., Duncan, S.H., Young, P., Belenguer, A., McWilliam Leitch, C., Scott, K.P., Flint, H.J., and Louis, P. (2014). Phylogenetic distribution of three pathways for propionate production within the human gut microbiota. ISME J. *8*, 1323–1335. https://doi.org/10.1038/ismej.2014.14.

Rémond, D., Shahar, D.R., Gille, D., Pinto, P., Kachal, J., Peyron, M.A., Dos Santos, C.N., Walther, B., Bordoni, A., Dupont, D., *et al.* (2015). Understanding the gastrointestinal tract of the elderly to develop dietary solutions that prevent malnutrition. Oncotarget *6*, 13858–13898.

Reuter, J.A., Spacek, D.V., and Snyder, M.P. (2015). High-throughput sequencing technologies. Mol. Cell *58*, 586–597. https://doi.org/10.1016/j.molcel.2015.05.004.

Reva, I.V., Yamamoto, T., Vershinina, S.S., and Reva, G.V. (2015). Immune homeostasis of human gastric mucosa in *Helicobacter pylori* infection. Bull. Exp. Biol. Med. *159*, 157–163. https://doi.org/10.1007/s10517-015-2913-x.

Reyes, A., Blanton, L.V., Cao, S., Zhao, G., Manary, M., Trehan, I., Smith, M.I., Wang, D., Virgin, H.W., Rohwer, F., *et al.* (2015). Gut DNA viromes of Malawian twins discordant for severe acute malnutrition. Proc. Natl. Acad. Sci. U.S.A. *112*, 11941–11946. https://doi.org/10.1073/pnas.1514285112.

Reyes, A., Haynes, M., Hanson, N., Angly, F.E., Heath, A.C., Rohwer, F., and Gordon, J.I. (2010). Viruses in the faecal microbiota of monozygotic twins and their mothers. Nature *466*, 334–338. https://doi.org/10.1038/nature09199.

Rios-Covian, D., Gueimonde, M., Duncan, S.H., Flint, H.J., and de los Reyes-Gavilan, C.G. (2015). Enhanced butyrate formation by cross-feeding between *Faecalibacterium prausnitzii* and *Bifidobacterium adolescentis*. FEMS Microbiol. Lett. *362*, fnv176. https://doi.org/10.1093/femsle/fnv176.

Ríos-Covián, D., Ruas-Madiedo, P., Margolles, A., Gueimonde, M., de Los Reyes-Gavilán, C.G., and Salazar, N. (2016). Intestinal short chain fatty acids and their link with diet and human health. Front. Microbiol. 7, 185. https://doi.org/10.3389/fmicb.2016.00185.

Roberfroid, M., Gibson, G.R., Hoyles, L., McCartney, A.L., Rastall, R., Rowland, I., Wolvers, D., Watzl, B., Szajewska, H., Stahl, B., *et al.* (2010). Prebiotic effects: Metabolic and health benefits. Br. J. Nutr. *104* (Suppl. 2), S1–63. https://doi.org/10.1017/S0007114510003363.

Rodríguez, J.M., Murphy, K., Stanton, C., Ross, R.P., Kober, O.I., Juge, N., Avershina, E., Rudi, K., Narbad, A., Jenmalm, M.C., *et al.* (2015). The composition of the gut microbiota throughout life, with an emphasis on early life. Microb. Ecol. Health Dis. *26*, 26050. https://doi.org/10.3402/mehd.v26.26050.

Rowe, J.W. (1997). The new gerontology. Science *278*, 367.

Salas-Salvadó, J., Rubio, M.A., Barbany, M., and Moreno, B. (2007). [SEEDO 2007 Consensus for the evaluation of overweight and obesity and the establishment of therapeutic intervention criteria.] Med. Clin. *128*, 184–196.

Salazar, N., Arboleya, S., Valdes, L., Stanton, C., Ross, P., Ruiz, L., Gueimonde, M., and de Los Reyes-Gavilán, C.G. (2014). The human intestinal microbiome at extreme ages of life: Dietary interventions as a way to counteract alterations. Front. Genet. *5*, 406. https://dx.doi.org/10.3389/fgene.2014.00406.

Salazar, N., López, P., Valdés, L., Margolles, A., Suárez, A., Patterson, A.M., Cuervo, A., de los Reyes-Gavilán, C.G., Ruas-Madiedo, P., Gonzalez, S., *et al.* (2013). Microbial targets for the development of functional foods accordingly with nutritional and immune parameters altered in the elderly. J. Am. Coll. Nutr. *32*, 399–406. https://doi.org/10.1080/07315724.2013.827047.

Salazar, N., Valdés-Varela, L., González, S., Gueimonde, M., and de Los Reyes-Gavilán, C.G. (2017). Nutrition and the gut microbiome in the elderly. Gut Microbes *8*, 82–97. https://doi.org/10.1080/19490976.2016.1256525.

Santiago-Rodriguez, T.M., Naidu, M., Abeles, S.R., Boehm, T.K., Ly, M., and Pride, D.T. (2015). Transcriptome analysis of bacteriophage communities in periodontal health and disease. BMC Genomics *16*, 549. https://doi.org/10.1186/s12864-015-1781-0.

Santoro, A., Ostan, R., Candela, M., Biagi, E., Brigidi, P., Capri, M., and Franceschi, C. (2018). Gut microbiota changes in the extreme decades of human life: A focus on centenarians. Cell Mol. Life Sci. *75*, 129–148. https://doi.org/10.1007/s00018-017-2674-y.

Scanlan, P.D., Shanahan, F., O'Mahony, C., and Marchesi, J.R. (2006). Culture-independent analyses of temporal variation of the dominant fecal microbiota and targeted bacterial subgroups in Crohn's disease. J. Clin. Microbiol. *44*, 3980–3988.

Sender, R., Fuchs, S., and Milo, R. (2016). Revised estimates for the number of human and bacteria cells in the body. PLOS Biol. *14*, e1002533. https://doi.org/10.1371/journal.pbio.1002533.

Sepehri, S., Kotlowski, R., Bernstein, C.N., and Krause, D.O. (2007). Microbial diversity of inflamed and noninflamed gut biopsy tissues in inflammatory bowel disease. Inflamm. Bowel Dis. *13*, 675–683. https://doi.org/10.1002/ibd.20101.

Shanahan, F., van Sinderen, D., O'Toole, P.W., and Stanton, C. (2017). Feeding the microbiota: Transducer of nutrient signals for the host. Gut *66*, 1709–1717. https://doi.org/10.1136/gutjnl-2017-313872.

Sharpton, T.J. (2014). An introduction to the analysis of shotgun metagenomic data. Front. Plant Sci. *5*, 209. https://doi.org/10.3389/fpls.2014.00209.

Shen, C.L., Cao, J.J., Dagda, R.Y., Chanjaplammootil, S., Lu, C., Chyu, M.C., Gao, W., Wang, J.S., and Yeh, J.K. (2012). Green tea polyphenols benefits body composition and improves bone quality in long-term high-fat diet-induced obese rats. Nutr. Res. *32*, 448–457. https://doi.org/10.1016/j.nutres.2012.05.001.

Smeeding, S.J. (2001). Nutrition, supplements, and aging. Geriatr. Nurs. *22*, 219–224.

Smirnov, K.S., Maier, T.V., Walker, A., Heinzmann, S.S., Forcisi, S., Martinez, I., Walter, J., and Schmitt-Kopplin, P. (2016). Challenges of metabolomics in human gut microbiota research. Int. J. Med. Microbiol. *306*, 266–279.

Soenen, S., Rayner, C.K., Jones, K.L., and Horowitz, M. (2016). The ageing gastrointestinal tract. Curr. Opin. Clin. Nutr. Metab. Care *19*, 12–18. https://doi.org/10.1097/MCO.0000000000000238.

Sonnenburg, E.D., Smits, S.A., Tikhonov, M., Higginbottom, S.K., Wingreen, N.S., and Sonnenburg, J.L. (2016). Diet-induced extinctions in the gut microbiota compound over generations. Nature *529*, 212–215. https://doi.org/10.1038/nature16504.

Sonnenburg, J.L., Xu, J., Leip, D.D., Chen, C.H., Westover, B.P., Weatherford, J., Buhler, J.D., and Gordon, J.I. (2005). Glycan foraging *in vivo* by an intestine-adapted bacterial symbiont. Science *307*, 1955–1959. https://doi.org/10.1126/science.1109051.

Spaiser, S.J., Culpepper, T., Nieves, C., Ukhanova, M., Mai, V., Percival, S.S., Christman, M.C., and Langkamp-Henken, B. (2015). *Lactobacillus gasseri* KS-13, *Bifidobacterium bifidum* G9-1, and *Bifidobacterium longum* MM-2 ingestion induces a less inflammatory cytokine profile and a potentially beneficial shift in gut microbiota in older adults: A randomized, double-blind, placebo-controlled, crossover study. J. Am. Coll. Nutr. *34*, 459–469. https://doi.org/10.1080/07315724.2014.983249.

Tarnawski, A.S., Ahluwalia, A., and Jones, M.K. (2014). Increased susceptibility of aging gastric mucosa to injury: the mechanisms and clinical implications. World J. Gastroenterol. *20*, 4467–4482. https://doi.org/10.3748/wjg.v20.i16.4467.

Tiihonen, K., Ouwehand, A.C., and Rautonen, N. (2010). Human intestinal microbiota and healthy ageing. Ageing Res. Rev. *9*, 107–116. https://doi.org/10.1016/j.arr.2009.10.004.

Tojo, R., Suárez, A., Clemente, M.G., de los Reyes-Gavilán, C.G., Margolles, A., Gueimonde, M., and Ruas-Madiedo, P. (2014). Intestinal microbiota in health and disease: Role of bifidobacteria in gut homeostasis. World J. Gastroenterol. *20*, 15163–15176. https://doi.org/10.3748/wjg.v20.i41.15163.

Tomb, J.F., White, O., Kerlavage, A.R., Clayton, R.A., Sutton, G.G., Fleischmann, R.D., Ketchum, K.A., Klenk, H.P., Gill, S., Dougherty, B.A., *et al.* (1997). The complete genome sequence of the gastric pathogen *Helicobacter pylori*. Nature *388*, 539–547. https://doi.org/10.1038/41483.

Topping, D.L., and Clifton, P.M. (2001). Short-chain fatty acids and human colonic function: Roles of resistant starch and nonstarch polysaccharides. Physiol. Rev. *81*, 1031–1064. https://doi.org/10.1152/physrev.2001.81.3.1031.

Troen, B.R. (2003). The biology of aging. Mt. Sinai J. Med. *70*, 3–22.

Tropini, C., Earle, K.A., Huang, K.C., and Sonnenburg, J.L. (2017). The gut microbiome: Connecting spatial organization to function. Cell Host Microbe *21*, 433–442. https://doi.org/10.1016/j.chom.2017.03.010.

Trumbo, P., Schlicker, S., Yates, A.A., and Poos, M. (2002). Dietary reference intakes for energy, carbohydrate, fiber, fat, fatty acids, cholesterol, protein and amino acids. J. Am. Diet. Assoc. *102*, 1621–1630.

Turnbaugh, P.J., Hamady, M., Yatsunenko, T., Cantarel, B.L., Duncan, A., Ley, R.E., Sogin, M.L., Jones, W.J., Roe, B.A., Affourtit, J.P., *et al.* (2009a). A core gut microbiome in obese and lean twins. Nature *457*, 480–484. https://doi.org/10.1038/nature07540.

Turnbaugh, P.J., Ridaura, V.K., Faith, J.J., Rey, F.E., Knight, R., and Gordon, J.I. (2009b). The effect of diet on the human gut microbiome: A metagenomic analysis in humanized gnotobiotic mice. Sci. Transl. Med. *1*, 6ra14. https://doi.org/10.1126/scitranslmed.3000322.

United Nations, Department of Economic and Social Affairs, Population Division (2015). World Population Prospects: The 2015 Revision, Key Findings and Advance Tables. Working Paper No. ESA/P//WP.241.

Ursell, L.K., Haiser, H.J., Van Treuren, W., Garg, N., Reddivari, L., Vanamala, J., Dorrestein, P.C., Turnbaugh, P.J., and Knight, R. (2014). The intestinal metabolome: An intersection between microbiota and host. Gastroenterology *146*, 1470–1476. https://doi.org/10.1053/j.gastro.2014.03.001.

Valdés, L., Salazar, N., González, S., Arboleya, S., Ríos-Covián, D., Genovés, S., Ramón, D., de los Reyes-Gavilán, C.G., Ruas-Madiedo, P., and Gueimonde, M. (2017). Selection of potential probiotic bifidobacteria and prebiotics for elderly by using *in vitro* faecal batch cultures. Eur. Food Res. Technol. *243*, 157–165. https://dx.doi.org/10.1007/s00217-016-2732-y.

Valdés-Varela, L., Alonso-Guervos, M., García-Suárez, O., Gueimonde, M., and Ruas-Madiedo, P. (2016). Screening of bifidobacteria and Lactobacilli able to antagonize the cytotoxic effect of *Clostridium difficile* upon intestinal epithelial HT29 monolayer. Front. Microbiol. *7*, 577. https://doi.org/10.3389/fmicb.2016.00577.

Vallejo, A.N. (2011). Immunological hurdles of ageing: Indispensable research of the human model. Ageing Res. Rev. *10*, 315–318. https://doi.org/10.1016/j.arr.2011.01.005.

Vendrame, S., Guglielmetti, S., Riso, P., Arioli, S., Klimis-Zacas, D., and Porrini, M. (2011). Six-week consumption of a wild blueberry powder drink increases bifidobacteria in the human gut. J. Agric. Food Chem. *59*, 12815–12820. https://doi.org/10.1021/jf2028686.

Vernocchi, P., Del Chierico, F., and Putignani, L. (2016). Gut microbiota profiling: Metabolomics based approach to unravel compounds affecting human health. Front. Microbiol. *7*, 1144. https://doi.org/10.3389/fmicb.2016.01144.

Vitali, B., Ndagijimana, M., Cruciani, F., Carnevali, P., Candela, M., Guerzoni, M.E., and Brigidi, P. (2010). Impact of a synbiotic food on the gut microbial ecology and metabolic profiles. BMC Microbiol. *10*, 4. https://doi.org/10.1186/1471-2180-10-4.

Vulevic, J., Drakoularakou, A., Yaqoob, P., Tzortzis, G., and Gibson, G.R. (2008). Modulation of the fecal microflora profile and immune function by a novel trans-galactooligosaccharide mixture (B-GOS) in healthy elderly volunteers. Am. J. Clin. Nutr. *88*, 1438–1446.

Vulevic, J., Juric, A., Walton, G.E., Claus, S.P., Tzortzis, G., Toward, R.E., and Gibson, G.R. (2015). Influence of galacto-oligosaccharide mixture (B-GOS) on gut microbiota, immune parameters and metabonomics in elderly persons. Br. J. Nutr. *114*, 586–595. https://doi.org/10.1017/S0007114515001889.

Walker, A.W. (2016). Studying the human microbiota. Adv. Exp. Med. Biol. *902*, 5–32. https://doi.org/10.1007/978-3-319-31248-4_2.

Walker, A.W., Duncan, S.H., Louis, P., and Flint, H.J. (2014). Phylogeny, culturing, and metagenomics of the human gut microbiota. Trends Microbiol. *22*, 267–274. https://doi.org/10.1016/j.tim.2014.03.001.

Wang, F., Yu, T., Huang, G., Cai, D., Liang, X., Su, H., Zhu, Z., Li, D., Yang, Y., Shen, P., *et al.* (2015). Gut microbiota community and its assembly associated with age and diet in Chinese centenarians. J. Microbiol. Biotechnol. *25*, 1195–1204. https://doi.org/10.4014/jmb.1410.10014.

Wang, Z., Gerstein, M., and Snyder, M. (2009). RNA-Seq: A revolutionary tool for transcriptomics. Nat. Rev. Genet. *10*, 57–63. https://doi.org/10.1038/nrg2484.

Willing, B.P., Russell, S.L., and Finlay, B.B. (2011). Shifting the balance: Antibiotic effects on host-microbiota mutualism. Nat. Rev. Microbiol. *9*, 233–243. https://doi.org/10.1038/nrmicro2536.

Wilmes, P., and Bond, P.L. (2009). Microbial community proteomics: Elucidating the catalysts and metabolic mechanisms that drive the Earth's biogeochemical cycles. Curr. Opin. Microbiol. *12*, 310–317. https://doi.org/10.1016/j.mib.2009.03.004.

Woese, C.R., and Fox, G.E. (1977). Phylogenetic structure of the prokaryotic domain: The primary kingdoms. Proc. Natl. Acad. Sci. U.S.A. *74*, 5088–5090.

Woese, C.R., Kandler, O., and Wheelis, M.L. (1990). Towards a natural system of organisms: Proposal for the domains Archaea, Bacteria, and Eucarya. Proc. Natl. Acad. Sci. U.S.A. *87*, 4576–4579.

Woodmansey, E.J. (2007). Intestinal bacteria and ageing. J. Appl. Microbiol. *102*, 1178–1186.

Woodmansey, E.J., McMurdo, M.E., Macfarlane, G.T., and Macfarlane, S. (2004). Comparison of compositions and metabolic activities of fecal microbiotas in young adults and in antibiotic-treated and non-antibiotic-treated elderly subjects. Appl. Environ. Microbiol. *70*, 6113–6122.

Wu, G.D., Chen, J., Hoffmann, C., Bittinger, K., Chen, Y.Y., Keilbaugh, S.A., Bewtra, M., Knights, D., Walters, W.A., Knight, R., *et al.* (2011). Linking long-term dietary patterns with gut microbial enterotypes. Science *334*, 105–108. https://doi.org/10.1126/science.1208344.

Wyss-Coray, T. (2016). Ageing, neurodegeneration and brain rejuvenation. Nature *539*, 180–186. https://doi.org/10.1038/nature20411.

Yang, J., Martínez, I., Walter, J., Keshavarzian, A., and Rose, D.J. (2013). *In vitro* characterization of the impact of selected dietary fibers on fecal microbiota composition and short chain fatty acid production. Anaerobe 23, 74–81. https://doi.org/10.1016/j.anaerobe.2013.06.012.

Yatsunenko, T., Rey, F.E., Manary, M.J., Trehan, I., Dominguez-Bello, M.G., Contreras, M., Magris, M., Hidalgo, G., Baldassano, R.N., Anokhin, A.P., *et al.* (2012). Human gut microbiome viewed across age and geography. Nature *486*, 222–227. https://doi.org/10.1038/nature11053.

Yen, C.H., Kuo, Y.W., Tseng, Y.H., Lee, M.C., and Chen, H.L. (2011). Beneficial effects of fructo-oligosaccharides supplementation on fecal bifidobacteria and index of peroxidation status in constipated nursing-home residents: A placebo-controlled, diet-controlled trial. Nutrition 27, 323–328. https://dx.doi.org/10.1016/j.nut.2010.02.009.

Yi, J., Wahl, K., Sederdahl, B.K., Jerris, R.R., Kraft, C.S., McCracken, C., Gillespie, S., Anderson, E.J., Kirby, A.E., Shane, A.L., *et al.* (2016). Molecular epidemiology of norovirus in children and the elderly in Atlanta, Georgia, United States. J. Med. Virol. *88*, 961–970. https://doi.org/10.1002/jmv.24436.

Zarowitz, B.J., Allen, C., O'Shea, T., and Strauss, M.E. (2015). Risk factors, clinical characteristics, and treatment differences between residents with and without nursing home- and non-nursing home-acquired *Clostridium difficile* infection. J. Manag. Care Spec. Pharm. *21*, 585–595.

Zeller, G., Tap, J., Voigt, A.Y., Sunagawa, S., Kultima, J.R., Costea, P.I., Amiot, A., Böhm, J., Brunetti, F., Habermann, N., *et al.* (2014). Potential of fecal microbiota for early-stage detection of colorectal cancer. Mol. Syst. Biol. *10*, 766. https://doi.org/10.15252/msb.20145645.

Zhang, C., and Zhao, L. (2016). Strain-level dissection of the contribution of the gut microbiome to human metabolic disease. Genome Med. *8*, 41. https://doi.org/10.1186/s13073-016-0304-1.

Zheng, D.P., Ando, T., Fankhauser, R.L., Beard, R.S., Glass, R.I., and Monroe, S.S. (2006). Norovirus classification and proposed strain nomenclature. Virology *346*, 312–323.

Zwielehner, J., Liszt, K., Handschur, M., Lassl, C., Lapin, A., and Haslberger, A.G. (2009). Combined PCR-DGGE fingerprinting and quantitative-PCR indicates shifts in fecal population sizes and diversity of *Bacteroides*, bifidobacteria and *Clostridium* cluster IV in institutionalized elderly. Exp. Gerontol. *44*, 440–446. https://doi.org/10.1016/j.exger.2009.04.002

Dysbiosis in Infants is an Important Environmental Factor of Developmental Origins of Health and Disease

2

Yuichiro Yamashiro[1*], Ravinder Nagpal[2], Hiromichi Shoji[3], Naoto Nishizaki[3] and Satoru Nagata[1,4]

[1]Probiotics Research Laboratory, Juntendo University Graduate School of Medicine, Tokyo, Japan.
[2]Centre for Diabetes, Obesity and Metabolism, Wake Forest School of Medicine, Winston-Salem, NC, USA.
[3]Department of Pediatrics and Adolescent Medicine, Juntendo University, Tokyo, Japan.
[4]Department of Pediatrics, Tokyo Women's Medical University, Tokyo, Japan.

*Correspondence: yamasiro@juntendo.ac.jp

https://doi.org/10.21775/9781910190937.02

Abstract

The original concept of the developmental origins of health and disease (DOHaD) was 'fetal origins of adult disease' but it is now revised as that 'early life environmental factors can have lifelong effects, and can manifest as disease in later life', based on the subsequent various studies. Recent studies demonstrated that gut microbiota plays an important role in human health and disease development. In particular, the aberrant bacterial community (dysbiosis) in early life can lead to diseases, such as non-communicable diseases, through altered development of the immune system and metabolic conditions. Therefore, the microbiota should be considered as an important environmental factor for DOHaD.

Introduction

The original concept of the developmental origins of health and disease (DOHaD) proposed by Barker was 'fetal origins of adult disease'. Subsequently, accumulated evidence has resulted in the revised concept that 'environmental factors during the phase of developmental plasticity interact with genotypic variation to change the capacity of the organism to cope with its environment in later life' (Barker *et al.*, 1986; Barker, 1995). Briefly, 'early life environmental factors can have lifelong effects, and can manifest as diseases in later life' (Gluckman *et al.*, 2006). However, aberrant bacterial community (dysbiosis) early in life can lead to disease through an altered development of the immune system and metabolic conditions.

The microbiome in early life and childhood: a dynamic interplay with the host as an important factor in DOHaD

Human gut microbiota is a complex ecosystem consisting of a total of 10^{14} bacteria. Its genome, which is more than 100 times larger than the human genome, can be defined as the microbiome. Commensal microbiota plays an important role in human health by acting as a barrier against pathogens and their invasion with highly dynamic modality, exerting metabolic functions and stimulating the development of immune systems.

Various studies (Penders *et al.*, 2006; Putignani *et al.*, 2014) demonstrated that the environment before and during pregnancy, delivery mode and host genetics can shape the newborn microbiome at birth. Subsequently, postnatal factors such as antibiotic treatment, diet or environmental exposure to a variety of microorganisms further modulate the development of the infant's microbiome and immune function. Furthermore, DOHaD studies in recent years have shown that factors such as caesarean (C) section (Darmasseelane *et al.*, 2014) and a high-fat diet (HFD) (Bäckhed *et al.*, 2004), which alter bacterial communities in infants during childhood, increase the risk for several diseases, highlighting the importance of understanding the early life microbiome composition, which is one of the important factors in DOHaD.

Gut microbiota of neonates and infants

Recent studies disclosed that microbes are present in the placenta, amniotic fluid, fetal membrane, umbilical cord blood and meconium (Rautava *et al.*, 2012; Funkhouser *et al.*, 2013). We previously analysed the faecal microbiota of 166 healthy Japanese newborns periodically from day 1 after birth until 3 years of age (Tsuji *et al.*, 2012) by using reverse-transcription quantitative polymerase chain reaction (PCR; Matsuda *et al.*, 2007, 2009). Faecal pH and organic acid concentration were also examined. Colonization by both facultative anaerobes and strict anaerobes was confirmed in 95% of meconium samples tested. A *Bifidobacterium*-predominant microbiota was subsequently established in most breastfed infants by three months after birth. *Bifidobacterium breve, Bifidobacterium longum, Bifidobacterium infantis, Bifidobacterium catenulatum* group and *Bifidobacterium bifidum* were the species mainly detected, that is, breastfed infants were found to develop *Bifidobacterium*-dominated microbiota by 3 months of age (Fig. 2.1). Intergroup correlation analysis revealed that bifidobacteria population levels, but not levels of other strict anaerobe groups, were negatively correlated with those of Enterobacteriaceae from seven days until three months after birth. Faecal pH was maintained at about 6 until 6 months after birth and reached 6.6 at 3 years after birth. The initial concentration of faecal organic acids just after birth (19 μmol/g of faeces) increased until 3 years after birth to 111 μmol/g of faeces. Early start of feeding formula milk promoted colonization by obligate anaerobes such as the *Clostridium coccoides* group, *Clostridium leptum* subgroup, *Prevotella* and *Atopobium* cluster during the 3 months after birth. Population levels of bifidobacteria until 1 month after birth and those of the *Bacteroides fragilis* group until 6 months after birth were lower in infants delivered by C-section than in those delivered normally. The results suggested that both an earlier start of feeding formula milk and

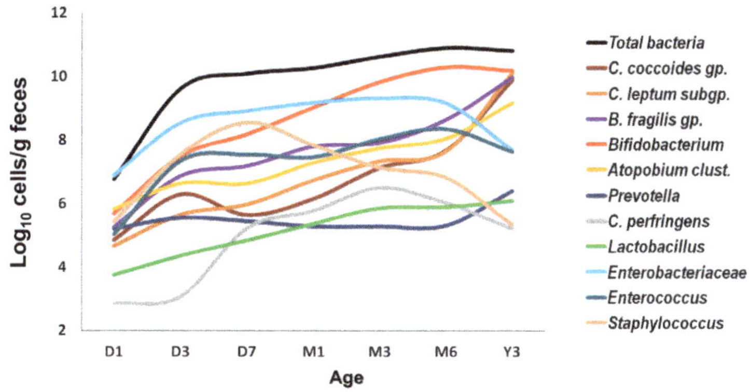

Figure 2.1 Gut microbiota profiles during the first three years of life. Colonization by both facultative and obligates anaerobes was confirmed in 95% of 151 meconium samples tested. *Bifidobacterium*-predominant microbiota was subsequently established in most breastfed infants by 3 months after birth.

the mode of infant delivery are important in the development of intestinal microbiota in early infancy.

Gut microbiota profiles of healthy children

The infant gut microbiota subsequently continues to evolve, exhibiting high levels of instability; this microbiota attains a stable and defined microbial pattern similar to that of an adult by two to three years of age. This process, however, can be influenced by a number of environmental and lifestyle factors such as eating habits, environmental hygiene, infection and medication (Ruemmele *et al.*, 2009). Our data show that an 'adult-type' intestinal microbiota profile may be established even in preschool children, that is, as early as four to five years of age (Wang *et al.*, 2015). Fig. 2.2 shows the main obligate anaerobes and facultative anaerobes in human throughout life from newborns to elders (> 80 years of age) based on our data (Nomoto *et al.*, 2015). *Bifidobacterium* is predominant until around adolescence but gradually decreases by 80 years of age, although it remains one of the main genera in the gut microbiota.

Factors influencing the early life gut microbiota composition

The early development and establishment of the gut microbiota during infancy is influenced by numerous factors (Fig. 2.3) (Putiguani *et al.*, 2014). Among these, mode of delivery, antibiotics and feeding remain the three foremost factors that play a prominent role in shaping the microbiota.

Delivery

The first major exposure of the newborn to microbes occurs during the birthing process and is highly dependent on the mode of delivery (Palmer *et al.*, 2007; Dominguez-Bello *et al.*, 2010; Bäckhed *et al.*, 2015). C-section-born infants acquire a different gut microbiota profile than vaginally born infants. The gut microbiota of vaginally born infants is enriched in *Lactobacillus* spp. and *Bifidobacterium*, which resembles the maternal vaginal microbiota

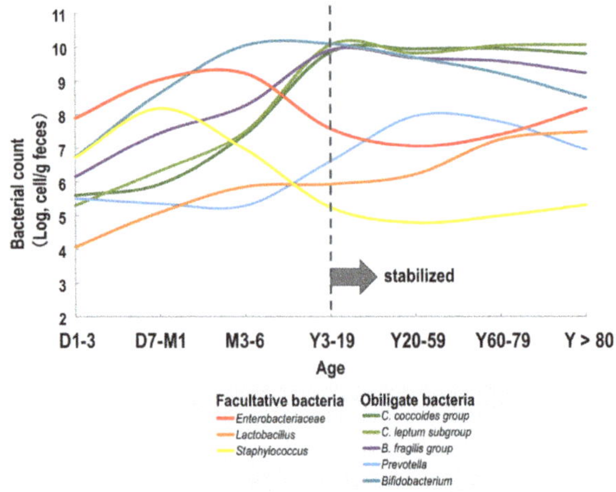

Figure 2.2 Gut microbiota profiles from newborns to elders. The main obligate and facultative anaerobes are shown. *Bifidobacterium* is predominant until around adolescence but gradually decreases by 80 years of age, although it remains one of the main genera in the gut microbiota. Source: Nomoto *et al*. (2015).

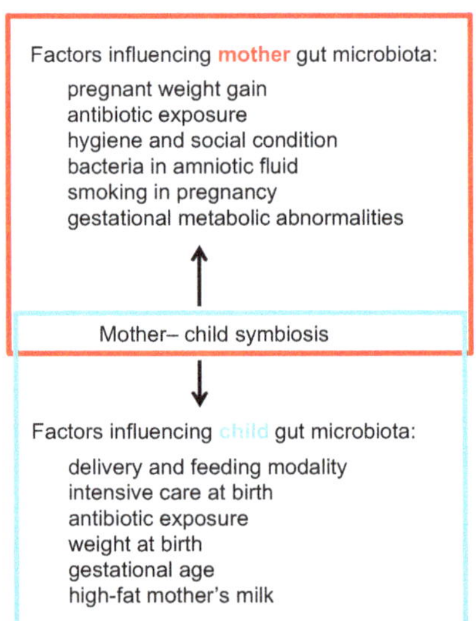

Figure 2.3 Mother–child symbiosis elements affecting the onset and modulation of the newborn gut microbiota. Adapted from Putignani *et al*. (2014).

(Fig. 2.2) (Bäckhed *et al.*, 2015). In contrast, infants delivered by C-section lack these microorganisms and are instead colonized by common skin and environmental microbes such as *Staphylococcus* and *Streptococcus*. In addition, we found that C-section-born infants had a higher carriage of α-toxigenic and enterotoxigenic *Clostridium perfringens* (Nagpal *et al.*, 2017). Such dissimilarities in the gut microbiota composition could still be detected at 19 years of age (Nagpal *et al.*, 2017). Dysbiosis is the prelude to a wide range of diseases, such as atopy and allergic diseases, obesity, type (T) 2 diabetes, T1 diabetes, metabolic diseases, inflammatory bowel diseases and behaviour abnormalities progressively appearing during childhood (Putiguani *et al.*, 2014).

Antibiotics

It is well recognized that the use of postnatal antibiotics can disrupt the delicate ecosystem of the neonatal microbiome. Most antibiotics are broad-spectrum antibiotics and do not target specific pathogens. Hence, in addition to killing targeted pathogen(s), these antibiotics can also influence the growth of commensal and beneficial gut microbes such as *Bacteroides* and *Escherichia coli*, as well as bifidobacteria (Savino *et al.*, 2011). The continuous use of antibiotics early in life has been linked to an increased risk of various conditions, similar to those mentioned above for delivery (Tamburini *et al.*, 2016).

Diet

Exclusively breastfed infants are found to harbour a significantly different gut microbiota from formula-fed infants. This may be ascribed to human milk oligosaccharides, which may favour the proliferation of specific beneficial bacteria and hence can influence the establishment of microbiota in breastfed infants (Penders *et al.*, 2006; Roger *et al.*, 2010). Breast milk may also contain up to 10^7 bacterial cells/800 ml, including bacteria such as bifidobacteria, lactobacilli, staphylococci and streptococci (Martín *et al.*, 2007; Collado *et al.*, 2009). In comparison to breastfed infants, formula-fed infants were found to have a complex adult-like microbiota, which is primarily composed of facultative anaerobes such as bacteroides, clostridia, staphylococci, streptococci and enterobacteria (Bezirtzoglou *et al.*, 2011; Azad *et al.*, 2013; Fan *et al.*, 2014). Several specific intestinal bifidobacterial strains from the mother have been found to be transmitted to the infant's gut, advocating that each mother–child pair may have distinct family-specific strains and breastfeeding may contribute in some way to the transfer from mother to infant (Makino *et al.*, 2011). Breastfeeding is also associated with numerous benefits for the child, such as increased resistance to infections (WHO Collaborative Study Team on the Role of Breastfeeding on the Prevention of Infant Mortality, 2000), lower risk of obesity (Weng *et al.*, 2006) and decreased risk of allergies (Greer, *et al.*, 2008).

DOHaD and the non-communicable disease epidemic

Epidemiological studies across a wide range of countries and over many years have shown that early human development affects the risk of developing non-communicable diseases (NCDs) in later life (Gluckman *et al.*, 2006). NCDs, namely cardiovascular disease, chronic respiratory disease, allergy and some forms of cancer (thyroid, liver, kidney and ovarian cancer), are the leading causes of death and disability globally, exacting a heavy and growing toll on the physical health and economic security of all countries (Narayan *et al.*, 2010). This field is now recognized as DOHaD.

Early life factors modulating later life phenotype

External environmental factors that are transmitted from mother to fetus or directly experienced by the neonate are shown in Fig. 2.1 (Hanson *et al.*, 2011). Signals received by the developing embryo, fetus and infant induce adaptive responses that enable the development of phenotypic characteristics appropriate for the environment in which the offspring predicts it will live. These responses are brought about via the mechanisms of developmental plasticity, which include epigenetic processes (e.g. epigenetic mechanisms – molecular mechanisms – that enable developing organisms to produce disparate and stable cellular phenotypes from the same genotype).

Early life nutritional excess

The earlier DOHaD research revealed an inverse relationship between birthweight and susceptibility to adult diseases, namely the link between low birthweight and higher adult risk of hypertension, cardiovascular morbidity, T2 diabetes, obesity and hyperlipidaemia. Subsequent work has, therefore, strongly focused on how maternal undernutrition affects offspring phenotype. However, the global prevalence of overweight, obesity and T2 diabetes has increased in adults and children in the past few decades. This situation has also led to increased numbers of women with gestational obesity and gestational diabetes mellitus, which are risk factors for the development of overweight, obesity and metabolic syndrome in their offspring (Poston *et al.*, 2011). Infants of overweight mothers have faster weight gain during the fetal period (i.e. higher birthweight) and infancy. It is now well recognized that early life overnutrition promotes a similar outcome to that of undernutrition, although the molecular and cellular mechanisms are not clear.

Very low birthweight and catch-up growth leading to obesity and metabolic syndrome

Very-low-birthweight (VLBW; birthweight < 1500 g) infants have an elevated risk of metabolic syndrome including T2 diabetes, hypertension and obesity in later life. This may commence with altered growth during the infantile period related to suboptimal nutritional intake [e.g. a high-protein diet for the purpose of catch-up growth, which leads to a high insulin-like growth factor (IGF-1) concentration (Gregaire *et al.*, 1998)]. Moreover, VLBW can potentially result in an aberrant microbiota with altered metabolic capacity, resulting in a greater percentage of body fat and considerably less lean mass at term-corrected age (Lucas, 2005; Plegemann *et al.*, 2012).

DOHaD and dysbiosis

The notion of DOHaD has been widely accepted within the context of metabolic and cardiovascular disorder (Fig. 2.4). It has been suggested that prenatal programming may also be influenced by various factors including microbial contact (Rautava *et al.*, 2012), as shown in Fig. 2.1. Indeed, recent observations disclosed that contact with complex bacterial communities is clearly already initiated *in utero* (Funkhouser *et al.*, 2013), and the impact of maternal microbiota may be more important than previously believed, namely the intrauterine environment is not sterile. In fact, we detected bacteria in meconium samples from normal vaginally born infants (Tsuji *et al.*, 2012). Furthermore, other groups confirmed diverse microbial communities in the placenta, umbilical cord, amniotic fluid and meconium (Jiménez *et al.*, 2005; Moles *et al.*, 2013; Aagaard *et al.*, 2014). Multiple recent studies employing meconium as a proxy for *in utero* bacterial communities suggest that bacterial transmission from mother to fetus is a regular occurrence during human pregnancies (Jiménez *et al.*, 2008).

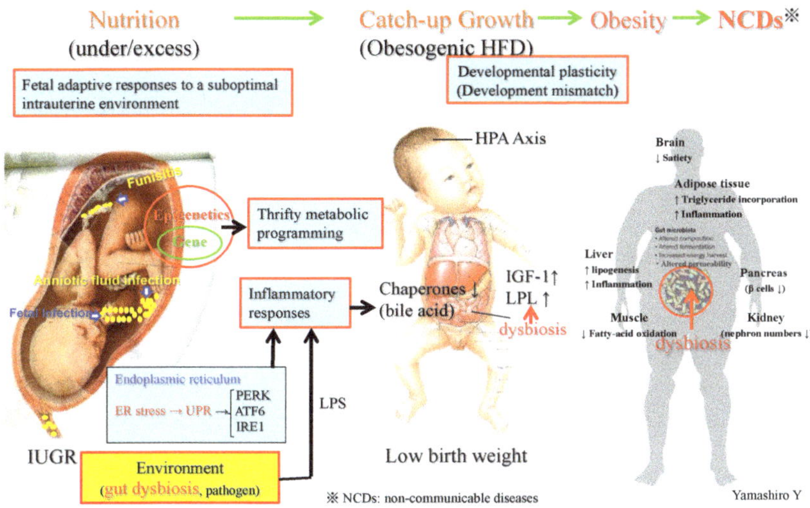

Figure 2.4 A conceptual framework for the DOHaD theory. The notion of DOHaD has been widely accepted within the context of metabolic and cardiovascular disorders, for which the microbiota should be considered an important environmental factor contributing to their development.

The fetal microbiome and immune development

Exposure to various microbiota antigens explains the relatively early development of a functional immune system in the human fetus. For example, $CD4^+$ and $CD8^+$ T cells can be identified at the end of the first trimester in the human fetus (Spencer *et al.*, 1986). Moreover, fetal T cells are highly responsive to stimulation and require control by an abundant and functional pool of Treg-specific forkhead transcription factor 3 ($FOXP3^+$) T regulatory cells.

Maternal mononuclear blood cells carry whole bacteria or their genetic material at a higher frequency and diversity during pregnancy (Perez *et al.*, 2007). Hence, *in utero* transfer of maternal microbial antigens, such as *Bifidobacteria* and *Lactobacillus*, during fetal development may enable a balanced immune response in the newborn to the rapidly developing microbiome post partum.

Maternal influence on fetal intestinal colonization

Manipulation of maternal nutrition during pre-pregnancy and pregnancy has been shown to directly impact fetal growth and development.

Maternal nutritional impact on the gut

A maternal high-fat diet influences the intestinal microbiome in mothers and their offspring (Ma *et al.*, 2014). In addition, a maternal high fat intake has been shown to impair fetal gut development, such as altered immune responses, and increased the Firmicutes-to-Bacteroidetes ratio, which is a well-known finding in obesity (Myles *et al.*, 2013).

Maternal microbiome

The maternal gut microbiota composition changes over the course of pregnancy. It is proposed that a lean woman possesses a stable, healthy gut microbiota that is modulated over the course of pregnancy, which may facilitate normal gut development and metabolic function in the offspring. Obese women likely present with disrupted gut microbiota before pregnancy, which is either amplified or further modified by pregnancy adaptations (Koren *et al.*, 2012).

Bacterial translocation from the intestine to the maternal bloodstream and subsequently to other organ systems is increased during pregnancy and lactation (Perez *et al.*, 2007). This process is active during human pregnancy and multiple pathways have been described for migration of luminal bacteria through the intact small intestinal epithelium, including dendritic cells (Macpherson *et al.*, 2004; Lelouard *et al.*, 2012), microfold cells (Jang *et al.*, 2004) and goblet cells (McDole *et al.*, 2012). Exposure of the fetal intestine to commensal organisms or their products in swallowed amniotic fluid may be an important contributor to gut maturation.

Maternal environment

During maternal viral or bacterial infections, immune stimulation can alter the intestinal microbiome during pregnancy, resulting in potentially long-lasting effects for the offspring, not only in the gastrointestinal system but also in distant organs including the brain (Hsiao *et al.*, 2013). Regulation of potentially adverse microbial metabolism after inflammatory exposure during pregnancy may be a powerful approach to reduce the risk of chronic diseases.

Maternal obesity

Based on extensive reviews, it is now clear that the gut microbiota is a contributor to weight gain and altered metabolism that accompanies obesity (Ley *et al.*, 2005; Turnbaugh *et al.*, 2009). Therefore, greater attention is being paid to the maternal gut microbiome and its association with maternal obesity and childhood outcome. Maternal obesity is currently one of the leading factors predicting childhood obesity (Patro *et al.*, 2013). Maternal obesity is associated with abnormal feto-placental function (Wallace *et al.*, 2012), offspring obesity risk (Patro *et al.*, 2013) and increased disease risk in general. Both maternal obesity and elevated gestational weight gain during pregnancy is associated with fetal macrosomia (Wallace *et al.*, 2012). Fetuses of obese mothers had a greater percentage of body fat and developed insulin resistance *in utero* (Catalano *et al.*, 2009). Furthermore, maternal body mass index is associated with activation of placental inflammatory pathways, and activation of the pro-inflammatory markers monocyte chemoattractant protein 1 (MCP-1) and tumour necrosis factor-alpha suggests a link between elevated pro-inflammatory cytokines in maternal plasma and activation of placental inflammatory pathways potentially influencing placental function (Aye *et al.*, 2014). Macrosomia is associated with an increased risk of developing obesity and metabolic syndrome (Sparano *et al.*, 2013), which appears to persist for multiple generations (Wostmann *et al.*, 1983).

Pathophysiological role of host microbiota in DOHaD

Diet–microbiota interaction

The human microbiota is affected by diet, and micronutrients serve as substrates for many microbially produced metabolites such as short-chain fatty acids (SCFAs) and bile acids (BAs) that may modulate host metabolism.

The diet–microbiota interaction is evident in that high fibre intake correlates with an increased abundance of *Prevotella*, whereas high fat and protein intake correlate with an increased abundance of *Bacteroides* (DeFilippo *et al.*, 2010; Yatsunenko *et al.*, 2012).

The gut microbiota is essential for efficient nutrient harvest from a polysaccharide-rich diet (Aiken *et al.*, 2014). Fermentation of polysaccharides in the human distal gut has been suggested to contribute to 10% of the daily energy requirement (McNeil *et al.*, 1984) and an increase in dietary fibre, leading to increased energy contribution (Iyayi *et al.*, 2015). Interestingly, fermentation of faeces from obese donors with different non-digestible carbohydrates was reported to produce higher cumulative amounts of SCFAs, which provide extra energy, compared with lean donors (Aguirre *et al.*, 2014).

Microbial signalling

Lipopolysaccharide and SCFAs

Cani *et al.* (2008) demonstrated that bacterial lipopolysaccharide (LPS), which is a powerful pro-inflammatory molecule and continuously produced in the gut through the lysis of Gram-negative bacteria, is a microbiota-related factor that can trigger increased gut permeability, low-grade inflammation and metabolic disorders in obesity and T2 diabetes (Saito *et al.*, 2007; Cani *et al.*, 2008). The relationship among fat intake, obesity, T2 diabetes and LPS has been confirmed in numerous studies (Erridge *et al.*, 2007; Creely *et al.*, 2007; Ghanim *et al.*, 2009; Pendyala *et al.*, 2012). We demonstrated that translocation of live bacteria to host tissue (bacteraemia) is accompanied by high serum LPS levels and dysbiosis. *Bifidobacterium* abundance appeared to be lower in overweight, obese or T2 diabetic patients than in lean subjects (Sato *et al.*, 2014). Furthermore, we identified a higher level of *Staphylococcus aureus* and lower levels of bifidobacteria in faecal samples from obese children, which may predict obesity and metabolic syndrome (Nagata *et al.*, 2017).

SCFAs act in the gut as signalling molecules and are specific ligands for at least two G-protein-coupled receptors, GPR41 and GPR43, mainly expressed in intestinal epithelial cells (Brown *et al.*, 2003), which affect peripheral glucose utilization (Samuel *et al.*, 2008).

Endocannabinoid system

The endocannabinoid (eCB) system is a complex of several bioactive lipids, enzymes and different types of receptors (Silvestri *et al.*, 2013) that plays a key role in the regulation of energy homeostasis and lipid and glucose metabolism at several levels (Cani *et al.*, 2014). Dysregulation of the eCB system leads to an unbalanced energy intake, contributes to excessive intraabdominal fat accumulation and is associated with the development of metabolic alterations observed in obesity and T2 diabetes. The gut microbiota determines adipose tissue physiology through LPS–eCB system regulatory loops and may have critical functions in adipose tissue plasticity during obesity.

Bile acids

Bile acids act as signalling molecules and activate nuclear BA receptors, called farnesoid X receptor (FXR) and G-protein-coupled receptor (TGR5), thereby regulating energy and hepatic lipid and glucose metabolism (Watanabe *et al.*, 2006; Thomas *et al.*, 2009). Stimulation of TGR5 attenuates diet-induced obesity (Silvestri *et al.*, 2013). TGR5 is expressed in L cells and its activation induces intestinal glucagon-like peptide-1 (GLP-1) release, leading to improved liver and endocrine pancreatic function and enhanced glucose tolerance in HFD mice. These data show that the TGR5 signalling pathway is critical in regulating intestinal GLP-1 secretion *in vivo* and suggest that pharmacological targeting of TGR5 may be a promising treatment strategy for metabolic disorders (Thomas *et al.*, 2008). BAs have bacteriostatic activity and a diet enriched in fats changes the BA composition, which influences the conditions for the gut microbial environment and causes dysbiosis (Turnbaugh *et al.*, 2012).

Improving dysbiosis to treat obesity and NCDs by probiotics (see Chapter 3)

Our treatment results, together with those of others, in preterm infants (Yamashiro *et al.*, 2010), obese children (Nagata *et al.*, 2017), T2 diabetic patients (Sato *et al.*, 2017) and elders (Nagata *et al.*, 2016) suggest that is may be possible to use probiotics to improve dysbiosis, leading to favourable metabolic features. Considering the establishment of the microbiota and presentation of dysbiosis can start from early life, even before pregnancy, strategies to prevent DOHaD are critical, such as appropriate nutritional status from preconception and pregnancy (fetus), breastfeeding (infants) and avoiding high-fat and high-protein diets (early childhood). Individualized treatment programmes based on the microbiota, if possible, may provide novel treatment strategies for DOHaD (Putignani *et al.*, 2014; Tamburiri *et al.*, 2016).

Conclusion

We propose that the microbiota should be considered as an important environmental factor that contributes to the development of obesity and obesity-related diseases, including metabolic diseases, namely DOHaD-associated diseases.

References

Aagaard, K., Ma, J., Antony, K.M., Ganu, R., Petrosino, J., and Versalovic, J. (2014). The placenta harbors a unique microbiome. Sci. Transl. Med. *6*, 237ra65. https://doi.org/10.1126/scitranslmed.3008599.

Aguirre, M., Jonkers, D.M., Troost, F.J., Roeselers, G., and Venema, K. (2014). *In vitro* characterization of the impact of different substrates on metabolite production, energy extraction and composition of gut microbiota from lean and obese subjects. PLOS ONE *9*, e113864. https://doi.org/10.1371/journal.pone.0113864.

Aiken, C.E., and Ozanne, S.E. (2014). Transgenerational developmental programming. Hum. Reprod. Update *20*, 63–75. https://doi.org/10.1093/humupd/dmt043.

Aye, I.L., Lager, S., Ramirez, V.I., Gaccioli, F., Dudley, D.J., Jansson, T., and Powell, T.L. (2014). Increasing maternal body mass index is associated with systemic inflammation in the mother and the activation of distinct placental inflammatory pathways. Biol. Reprod. *90*, 129. https://doi.org/10.1095/biolreprod.113.116186.

Azad, M.B., Konya, T., Maughan, H., Guttman, D.S., Field, C.J., Chari, R.S., Sears, M.R., Becker, A.B., Scott, J.A., and Kozyrskyj, A.L. (2013). Gut microbiota of healthy Canadian infants: Profiles by mode of delivery and infant diet at 4 months. CMAJ *185*, 385–394. https://doi.org/10.1503/cmaj.121189.

Bäckhed, F., Ding, H., Wang, T., Hooper, L.V., Koh, G.Y., Nagy, A., Semenkovich, C.F., and Gordon, J.I. (2004). The gut microbiota as an environmental factor that regulates fat storage. Proc. Natl. Acad. Sci. U.S.A. *101*, 15718–15723.

Bäckhed, F., Roswall, J., Peng, Y., Feng, Q., Jia, H., Kovatcheva-Datchary, P., Li, Y., Xia, Y., Xie, H., Zhong, H., *et al.* (2015). Dynamics and stabilization of the human gut microbiome during the first year of life. Cell Host Microbe *17*, 690–703. https://doi.org/10.1016/j.chom.2015.04.004.

Barker, D.J. (1995). Fetal origins of coronary heart disease. BMJ *311*, 171–174.

Barker, D.J., and Osmond, C. (1986). Infant mortality, childhood nutrition, and ischaemic heart disease in England and Wales. Lancet *1*, 1077–1081.

Bezirtzoglou, E., Tsiotsias, A., and Welling, G.W. (2011). Microbiota profile in feces of breast- and formula-fed newborns by using fluorescence *in situ* hybridization (FISH). Anaerobe *17*, 478–482. https://doi.org/10.1016/j.anaerobe.2011.03.009.

Brown, A.J., Goldsworthy, S.M., Barnes, A.A., Eilert, M.M., Tcheang, L., Daniels, D., Muir, A.I., Wigglesworth, M.J., Kinghorn, I., Fraser, N.J., *et al.* (2003). The Orphan G protein-coupled receptors GPR41 and GPR43 are activated by propionate and other short chain carboxylic acids. J. Biol. Chem. *278*, 11312–11319. https://doi.org/10.1074/jbc.M211609200.

Catalano, P.M., Presley, L., Minium, J., and Hauguel-de Mouzon, S. (2009). Fetuses of obese mothers develop insulin resistance *in utero*. Diabetes Care *32*, 1076–1080. https://doi.org/10.2337/dc08-2077.

Cani, P.D., Bibiloni, R., Knauf, C., Waget, A., Neyrinck, A.M., Delzenne, N.M., and Burcelin, R. (2008). Changes in gut microbiota control metabolic endotoxemia-induced inflammation in high-fat diet-induced obesity and diabetes in mice. Diabetes *57*, 1470–1481. https://doi.org/10.2337/db07-1403.

Cani, P.D., Geurts, L., Matamoros, S., Plovier, H., and Duparc, T. (2014). Glucose metabolism: Focus on gut microbiota, the endocannabinoid system and beyond. Diabetes Metab. *40*, 246–257. https://doi.org/10.1016/j.diabet.2014.02.004.

Collado, M.C., Delgado, S., Maldonado, A., and Rodríguez, J.M. (2009). Assessment of the bacterial diversity of breast milk of healthy women by quantitative real-time PCR. Lett. Appl. Microbiol. *48*, 523–528. https://doi.org/10.1111/j.1472-765X.2009.02567.x.

Creely, S.J., McTernan, P.G., Kusminski, C.M., Fisher, f.M., Da Silva, N.F., Khanolkar, M., Evans, M., Harte, A.L., and Kumar, S. (2007). Lipopolysaccharide activates an innate immune system response in human adipose tissue in obesity and type 2 diabetes. Am. J. Physiol. Endocrinol. Metab. *292*, E740–7.

Darmasseelane, K., Hyde, M.J., Santhakumaran, S., Gale, C., and Modi, N. (2014). Mode of delivery and offspring body mass index, overweight and obesity in adult life: A systematic review and meta-analysis. PLOS ONE *9*, e87896. https://doi.org/10.1371/journal.pone.0087896.

De Filippo, C., Cavalieri, D., Di Paola, M., Ramazzotti, M., Poullet, J.B., Massart, S., Collini, S., Pieraccini, G., and Lionetti, P. (2010). Impact of diet in shaping gut microbiota revealed by a comparative study in children from Europe and rural Africa. Proc. Natl. Acad. Sci. U.S.A. *107*, 14691–14696. https://doi.org/10.1073/pnas.1005963107.

Dominguez-Bello, M.G., Costello, E.K., Contreras, M., Magris, M., Hidalgo, G., Fierer, N., and Knight, R. (2010). Delivery mode shapes the acquisition and structure of the initial microbiota across multiple body habitats in newborns. Proc. Natl. Acad. Sci. U.S.A. *107*, 11971–11975. https://doi.org/10.1073/pnas.1002601107.

Erridge, C., Attina, T., Spickett, C.M., and Webb, D.J. (2007). A high-fat meal induces low-grade endotoxemia: Evidence of a novel mechanism of postprandial inflammation. Am. J. Clin. Nutr. *86*, 1286–1292.

Fan, W., Huo, G., Li, X., Yang, L., and Duan, C. (2014). Impact of diet in shaping gut microbiota revealed by a comparative study in infants during the six months of life. J. Microbiol. Biotechnol. *24*, 133–143.

Funkhouser, L.J., and Bordenstein, S.R. (2013). Mom knows best: The universality of maternal microbial transmission. PLOS Biol. *11*, e1001631. https://doi.org/10.1371/journal.pbio.1001631.

Ghanim, H., Abuaysheh, S., Sia, C.L., Korzeniewski, K., Chaudhuri, A., Fernandez-Real, J.M., and Dandona, P. (2009). Increase in plasma endotoxin concentrations and the expression of Toll-like receptors and suppressor of cytokine signaling-3 in mononuclear cells after a high-fat, high-carbohydrate meal: Implications for insulin resistance. Diabetes Care *32*, 2281–2287. https://doi.org/10.2337/dc09-0979.

Gluckman, P.D., and Hanson, M.A. (2006). The conceptual basis for the developmental origins of health and disease. In Developmental Origins of Health and Disease, Gluckman, P.D., and Hanson, M.A., eds. (Cambridge University Press, Cambridge), pp. 1–5.

Greer, F.R., Sicherer, S.H., and Burks, A.W. (2008). Effects of early nutritional interventions on the development of atopic disease in infants and children: The role of maternal dietary restriction,

breastfeeding, timing of introduction of complementary foods, and hydrolyzed formulas. Pediatrics *121*, 183–191. https://doi.org/10.1542/peds.2007-3022.
Gregoire, F.M., Smas, C.M., and Sul, H.S. (1998). Understanding adipocyte differentiation. Physiol. Rev. *78*, 783–809. https://doi.org/10.1152/physrev.1998.78.3.783.
Hanson, M.A., Low, F.M., and Gluckman, P.D. (2011). Epigenetic epidemiology: The rebirth of soft inheritance. Ann. Nutr. Metab. *58* (Suppl. 2), 8–15. https://doi.org/10.1159/000328033.
Hsiao, E.Y., McBride, S.W., Hsien, S., Sharon, G., Hyde, E.R., McCue, T., Codelli, J.A., Chow, J., Reisman, S.E., Petrosino, J.F., *et al.* (2013). Microbiota modulate behavioral and physiological abnormalities associated with neurodevelopmental disorders. Cell *155*, 1451–1463. https://doi.org/10.1016/j.cell.2013.11.024.
Iyayi, E.A., and Adeola, O. (2015). Quantification of short-chain fatty acids and energy production from hindgut fermentation in cannulated pigs fed graded levels of wheat bran. J. Anim. Sci. *93*, 4781–4787. https://doi.org/10.2527/jas.2015-9081.
Jang, M.H., Kweon, M.N., Iwatani, K., Yamamoto, M., Terahara, K., Sasakawa, C., Suzuki, T., Nochi, T., Yokota, Y., Rennert, P.D., *et al.* (2004). Intestinal villous M cells: An antigen entry site in the mucosal epithelium. Proc. Natl. Acad. Sci. U.S.A. *101*, 6110–6115. https://doi.org/10.1073/pnas.0400969101.
Jiménez, E., Fernández, L., Marín, M.L., Martín, R., Odriozola, J.M., Nueno-Palop, C., Narbad, A., Olivares, M., Xaus, J., and Rodríguez, J.M. (2005). Isolation of commensal bacteria from umbilical cord blood of healthy neonates born by cesarean section. Curr. Microbiol. *51*, 270–274. https://doi.org/10.1007/s00284-005-0020-3.
Jiménez, E., Marín, M.L., Martín, R., Odriozola, J.M., Olivares, M., Xaus, J., Fernández, L., and Rodríguez, J.M. (2008). Is meconium from healthy newborns actually sterile? Res. Microbiol. *159*, 187–193. https://doi.org/10.1016/j.resmic.2007.12.007.
Koren, O., Goodrich, J.K., Cullender, T.C., Spor, A., Laitinen, K., Bäckhed, H.K., Gonzalez, A., Werner, J.J., Angenent, L.T., Knight, R., *et al.* (2012). Host remodeling of the gut microbiome and metabolic changes during pregnancy. Cell *150*, 470–480. https://doi.org/10.1016/j.cell.2012.07.008.
Lelouard, H., Fallet, M., de Bovis, B.,Méresse, S., and Gorvel, J.P. (2012). Peyer's patch dendritic cells sample antigens by extending dendrites through M cell-specific transcellular pores. Gastroenterology *142*, 592–601.e3. https://doi.org/10.1053/j.gastro.2011.11.039.
Ley, R.E., Bäckhed, F., Turnbaugh, P., Lozupone, C.A., Knight, R.D., and Gordon, J.I. (2005). Obesity alters gut microbial ecology. Proc. Natl. Acad. Sci. U.S.A. *102*, 11070–11075.
Lucas, A. (2005). Long-term programming effects of early nutrition – implications for the preterm infant. J. Perinatol. *25* (Suppl. 2), S2–6.
Ma, J., Prince, A.L., Bader, D., Hu, M., Ganu, R., Baquero, K., Blundell, P., Alan Harris, R., Frias, A.E., Grove, K.L., *et al.* (2014). High-fat maternal diet during pregnancy persistently alters the offspring microbiome in a primate model. Nat. Commun. *5*, 3889. https://doi.org/10.1038/ncomms4889.
Macpherson, A.J., and Uhr, T. (2004). Induction of protective IgA by intestinal dendritic cells carrying commensal bacteria. Science *303*, 1662–1665. https://doi.org/10.1126/science.1091334.
Makino, H., Kushiro, A., Ishikawa, E., Muylaert, D., Kubota, H., Sakai, T., Oishi, K., Martin, R., Ben Amor, K., Oozeer, R., *et al.* (2011). Transmission of intestinal *Bifidobacterium longum* subsp. *longum* strains from mother to infant, determined by multilocus sequencing typing and amplified fragment length polymorphism. Appl. Environ. Microbiol. *77*, 6788–6793. https://doi.org/10.1128/AEM.05346-11.
Matsuda, K., Tsuji, H., Asahara, T., Kado, Y., and Nomoto, K. (2007). Sensitive quantitative detection of commensal bacteria by rRNA-targeted reverse transcription-PCR. Appl. Environ. Microbiol. *73*, 32–39.
Matsuda, K., Tsuji, H., Asahara, T., Matsumoto, K., Takada, T., and Nomoto, K. (2009). Establishment of an analytical system for the human fecal microbiota, based on reverse transcription-quantitative PCR targeting of multicopy rRNA molecules. Appl. Environ. Microbiol. *75*, 1961–1969. https://doi.org/10.1128/AEM.01843-08.
Martín, R., Heilig, H.G., Zoetendal, E.G., Jiménez, E., Fernández, L., Smidt, H., and Rodríguez, J.M. (2007). Cultivation-independent assessment of the bacterial diversity of breast milk among healthy women. Res. Microbiol. *158*, 31–37.
McDole, J.R., Wheeler, L.W., McDonald, K.G., Wang, B., Konjufca, V., Knoop, K.A., Newberry, R.D., and Miller, M.J. (2012). Goblet cells deliver luminal antigen to CD103+ dendritic cells in the small intestine. Nature *483*, 345–349. https://doi.org/10.1038/nature10863.
McNeil, N.I. (1984). The contribution of the large intestine to energy supplies in man. Am. J. Clin. Nutr. *39*, 338–342. https://doi.org/10.1093/ajcn/39.2.338.
Moles, L., Gómez, M., Heilig, H., Bustos, G., Fuentes, S., de Vos, W., Fernández, L., Rodríguez, J.M., and Jiménez, E. (2013). Bacterial diversity in meconium of preterm neonates and evolution of their fecal

microbiota during the first month of life. PLOS ONE *8*, e66986. https://doi.org/10.1371/journal.pone.0066986.

Myles, I.A., Fontecilla, N.M., Janelsins, B.M., Vithayathil, P.J., Segre, J.A., and Datta, S.K. (2013). Parental dietary fat intake alters offspring microbiome and immunity. J. Immunol. *191*, 3200–3209. https://doi.org/10.4049/jimmunol.1301057.

Nagata, S., Chiba, Y., Wang, C., and Yamashiro, Y. (2017). The effects of the *Lactobacillus casei* strain on obesity in children: A pilot study. Benef. Microbes *8*, 535–543. https://doi.org/10.3920/BM2016.0170.

Nagata, S., Asahara, T., Wang, C., Suyama, Y., Chonan, O., Takano, K., Daibou, M., Takahashi, T., Nomoto, K., and Yamashiro, Y. (2016). The effectiveness of *Lactobacillus* beverages in controlling infections among the residents of an aged care facility: A randomized placebo-controlled double-blind trial. Ann. Nutr. Metab. *68*, 51–59. https://doi.org/10.1159/000442305.

Nagpal, R., Tsuji, H., Takahashi, T., Nomoto, K., Kawashima, K., Nagata, S., and Yamashiro, Y. (2017). Gut dysbiosis following C-section instigates higher colonisation of toxigenic Clostridium perfringens in infants. Benef. Microbes *8*, 353–365. https://doi.org/10.3920/BM2016.0216.

Narayan, K.M., Ali, M.K., and Koplan, J.P. (2010). Global noncommunicable diseases – where worlds meet. N. Engl. J. Med. *363*, 1196–1198. https://doi.org/10.1056/NEJMp1002024.

Nomoto, K., and Tsuji, H. (2015). [Establishment of a novel intestinal microbiota analysis system (YIF-SCAN®) and its application]. J. JSMUFF. *8*, 532–537.

Palmer, C., Bik, E.M., DiGiulio, D.B., Relman, D.A., and Brown, P.O. (2007). Development of the human infant intestinal microbiota. PLOS Biol. *5*, e177.

Patro, B., Liber, A., Zalewski, B., Poston, L., Szajewska, H., and Koletzko, B. (2013). Maternal and paternal body mass index and offspring obesity: A systematic review. Ann. Nutr. Metab. *63*, 32–41. https://doi.org/10.1159/000350313.

Penders, J., Thijs, C., Vink, C., Stelma, F.F., Snijders, B., Kummeling, I., van den Brandt, P.A., and Stobberingh, E.E. (2006). Factors influencing the composition of the intestinal microbiota in early infancy. Pediatrics *118*, 511–521.

Pendyala, S., Walker, J.M., and Holt, P.R. (2012). A high-fat diet is associated with endotoxemia that originates from the gut. Gastroenterology *142*, 1100–1101.e2. https://doi.org/10.1053/j.gastro.2012.01.034.

Perez, P.F., Doré, J., Leclerc, M., Levenez, F., Benyacoub, J., Serrant, P., Segura-Roggero, I., Schiffrin, E.J., and Donnet-Hughes, A. (2007). Bacterial imprinting of the neonatal immune system: Lessons from maternal cells? Pediatrics *119*, e724–32.

Plagemann, A., Harder, T., Schellong, K., Schulz, S., and Stupin, J.H. (2012). Early postnatal life as a critical time window for determination of long-term metabolic health. Best Pract. Res. Clin. Endocrinol. Metab. *26*, 641–653. https://doi.org/10.1016/j.beem.2012.03.008.

Poston, L., Harthoorn, L.F., and Van Der Beek, E.M. (2011). Obesity in pregnancy: Implications for the mother and lifelong health of the child. A consensus statement. Pediatr. Res. *69*, 175–180. https://doi.org/10.1203/PDR.0b013e3182055ede.

Putignani, L., Del Chierico, F., Petrucca, A., Vernocchi, P., and Dallapiccola, B. (2014). The human gut microbiota: A dynamic interplay with the host from birth to senescence settled during childhood. Pediatr. Res. *76*, 2–10. https://doi.org/10.1038/pr.2014.49.

Rautava, S., Luoto, R., Salminen, S., and Isolauri, E. (2012). Microbial contact during pregnancy, intestinal colonization and human disease. Nat. Rev. Gastroenterol. Hepatol. *9*, 565–576. https://doi.org/10.1038/nrgastro.2012.144.

Roger, L.C., and McCartney, A.L. (2010). Longitudinal investigation of the faecal microbiota of healthy full-term infants using fluorescence *in situ* hybridization and denaturing gradient gel electrophoresis. Microbiology *156*, 3317–3328. https://doi.org/10.1099/mic.0.041913-0.

Ruemmele, F.M., Bier, D., Marteau, P., Rechkemmer, G., Bourdet-Sicard, R., Walker, W.A., and Goulet, O. (2009). Clinical evidence for immunomodulatory effects of probiotic bacteria. J. Pediatr. Gastroenterol. Nutr. *48*, 126–141. https://doi.org/10.1097/MPG.0b013e31817d80ca.

Saito, T., Hayashida, H., and Furugen, R. (2007). Comment on: Cani *et al.* (2007) Metabolic endotoxemia initiates obesity and insulin resistance: Diabetes 56:1761-1772. Diabetes *56*, e20.

Samuel, B.S., Shaito, A., Motoike, T., Rey, F.E., Backhed, F., Manchester, J.K., Hammer, R.E., Williams, S.C., Crowley, J., Yanagisawa, M., *et al.* (2008). Effects of the gut microbiota on host adiposity are modulated by the short-chain fatty-acid binding G protein-coupled receptor, Gpr41. Proc. Natl. Acad. Sci. U.S.A. *105*, 16767–16772. https://doi.org/10.1073/pnas.0808567105.

Sato, J., Kanazawa, A., Azuma, K., Ikeda, F., Goto, H., Komiya, K., Kanno, R., Tamura, Y., Asahara, T., Takahashi, T., *et al.* (2017). Probiotic reduces bacterial translocation in type 2 diabetes mellitus: A randomised controlled study. Sci. Rep. 7, 12115. https://doi.org/10.1038/s41598-017-12535-9.

Sato, J., Kanazawa, A., Ikeda, F., Yoshihara, T., Goto, H., Abe, H., Komiya, K., Kawaguchi, M., Shimizu, T., Ogihara, T., *et al.* (2014). Gut dysbiosis and detection of 'live gut bacteria' in blood of Japanese patients with type 2 diabetes. Diabetes Care 37, 2343–2350. https://doi.org/10.2337/dc13-2817.

Savino, F., Roana, J., Mandras, N., Tarasco, V., Locatelli, E., and Tullio, V. (2011). Faecal microbiota in breast-fed infants after antibiotic therapy. Acta Paediatr. *100*, 75–78. https://doi.org/10.1111/j.1651-2227.2010.01988.x.

Silvestri, C., and Di Marzo, V. (2013). The endocannabinoid system in energy homeostasis and the etiopathology of metabolic disorders. Cell Metab. *17*, 475–490. https://doi.org/10.1016/j.cmet.2013.03.001.

Sparano, S., Ahrens, W., De Henauw, S., Marild, S., Molnar, D., Moreno, L.A., Suling, M., Tornaritis, M., Veidebaum, T., Siani, A., *et al.* (2013). Being macrosomic at birth is an independent predictor of overweight in children: Results from the IDEFICS study. Matern. Child Health J. *17*, 1373–1381. https://doi.org/10.1007/s10995-012-1136-2.

Spencer, J., Dillon, S.B., Isaacson, P.G., and MacDonald, T.T. (1986). T cell subclasses in fetal human ileum. Clin. Exp. Immunol. *65*, 553–558.

Tamburini, S., Shen, N., Wu, H.C., and Clemente, J.C. (2016). The microbiome in early life: Implications for health outcomes. Nat. Med. *22*, 713–722. https://doi.org/10.1038/nm.4142.

Thomas, C., Pellicciari, R., Pruzanski, M., Auwerx, J., and Schoonjans, K. (2008). Targeting bile-acid signalling for metabolic diseases. Nat. Rev. Drug Discov. *7*, 678–693. https://doi.org/10.1038/nrd2619.

Thomas, C., Gioiello, A., Noriega, L., Strehle, A., Oury, J., Rizzo, G., Macchiarulo, A., Yamamoto, H., Mataki, C., Pruzanski, M., *et al.* (2009). TGR5-mediated bile acid sensing controls glucose homeostasis. Cell Metab. *10*, 167–177. https://doi.org/10.1016/j.cmet.2009.08.001.

Tsuji, H., Oozeer, R., Matsuda, K., Matsuki, T., Ohta, T., Nomoto, K., Tanaka, R., Kawashima, M., Kawashima, K., Nagata, S., *et al.* (2012). Molecular monitoring of the development of intestinal microbiota in Japanese infants. Benef. Microbes *3*, 113–125. https://doi.org/10.3920/BM2011.0038.

Turnbaugh, P.J., Hamady, M., Yatsunenko, T., Cantarel, B.L., Duncan, A., Ley, R.E., Sogin, M.L., Jones, W.J., Roe, B.A., Affourtit, J.P., *et al.* (2009). A core gut microbiome in obese and lean twins. Nature *457*, 480–484. https://doi.org/10.1038/nature07540.

Turnbaugh, P.J. (2012). Microbiology: Fat, bile and gut microbes. Nature *487*, 47–48. https://doi.org/10.1038/487047a.

Wallace, J.M., Horgan, G.W., and Bhattacharya, S. (2012). Placental weight and efficiency in relation to maternal body mass index and the risk of pregnancy complications in women delivering singleton babies. Placenta *33*, 611–618. https://doi.org/10.1016/j.placenta.2012.05.006.

Watanabe, M., Houten, S.M., Mataki, C., Christoffolete, M.A., Kim, B.W., Sato, H., Messaddeq, N., Harney, J.W., Ezaki, O., Kodama, T., *et al.* (2006). Bile acids induce energy expenditure by promoting intracellular thyroid hormone activation. Nature *439*, 484–489.

Wang, C., Nagata, S., Asahara, T., Yuki, N., Matsuda, K., Tsuji, H., Takahashi, T., Nomoto, K., and Yamashiro, Y. (2015). Intestinal microbiota profiles of healthy pre-school and school-age children and effects of probiotic supplementation. Ann. Nutr. Metab. *67*, 257–266. https://doi.org/10.1159/000441066.

Weng, S.F., Redsell, S.A., Swift, J.A., Yang, M., and Glazebrook, C.P. (2012). Systematic review and meta-analyses of risk factors for childhood overweight identifiable during infancy. Arch. Dis. Child. *97*, 1019–1026. https://doi.org/10.1136/archdischild-2012-302263.

WHO Collaborative Study Team on the Role of Breastfeeding on the Prevention of Infant Mortality. (2000). Effect of breastfeeding on infant and child mortality due to infectious diseases in less developed countries: A pooled analysis. Lancet 355, 451–455.

Wostmann, B.S., Larkin, C., Moriarty, A., and Bruckner-Kardoss, E. (1983). Dietary intake, energy metabolism, and excretory losses of adult male germfree Wistar rats. Lab. Anim. Sci. 33, 46–50.

Yamashiro, Y., and Nagata, S. (2010). Beneficial microbes for premature infants, and children with malignancy undergoing chemotherapy. Benef. Microbes *1*, 357–365. https://doi.org/10.3920/BM2010.0035.

Yatsunenko, T., Rey, F.E., Manary, M.J., Trehan, I., Dominguez-Bello, M.G., Contreras, M., Magris, M., Hidalgo, G., Baldassano, R.N., Anokhin, A.P., *et al.* (2012). Human gut microbiome viewed across age and geography. Nature *486*, 222–227. https://doi.org/10.1038/nature11053

Beneficial Effects of Probiotics for Infants and Children with Dysbiosis

3

Yuichiro Yamashiro[1*], Hiromichi Shoji[2], Naoto Nishizaki[3] and Satoru Nagata[1,4]

[1]Probiotics Research Laboratory, Juntendo University Graduate School of Medicine, Tokyo, Japan.
[2]Department of Pediatrics and Adolescent Medicine, Juntendo University, Tokyo, Japan.
[3]Department of Pediatrics, Juntendo University Urayasu Hospital, Urayasu, Japan.
[4]Department of Pediatrics, Tokyo Women's Medical University, Tokyo, Japan.

*Correspondence: yamasiro@juntendo.ac.jp

https://doi.org/10.21775/9781910190937.03

Abstract

Epidemiological studies have established a clear correlation between factors that disrupt the intestinal bacterial community (dysbiosis) early in life and diseases and metabolic conditions associated with altered development of the immune system.

Recent studies have also described interventions that modify host microbial composition, suggesting that overcoming microbial imbalances during childhood may serve as a preventative therapeutic approach. Probiotics are defined as live microorganisms that, when taken in a sufficient amount, have a benefit for the ingesting host. The use of probiotics is one approach to preventing and regulating dysbiosis in infants and children. The biomedical literature has recently featured an increasing number of prospective, randomized controlled trials indicating that probiotics can be employed as an effective strategy for the treatment of a variety of diseases affecting children. This review will focus on systematic reviews highlighting the efficacy of probiotics in treating paediatric illnesses such as pre-term low-birth-weight infants with necrotizing enterocolitis and infection, infants born by caesarean section with allergy, children with mucositis and children with obesity. Our own trial testing a probiotic supplement for use by healthy children will also be described.

Introduction

Today, obesity has become a pandemic. Overweight and obesity increase the risk for cardiovascular diseases, metabolic syndromes such as type 2 diabetes mellitus (T2DM), dyslipidaemia and hepatobiliary disease, and lung, breast, uterine, and ovarian cancer. The US Centers for Disease Control and Prevention recently (October 2017) released a report on cancer and obesity indicating that cancers associated with overweight and obesity constitute 40% of cancers diagnosed in the USA (*The Lancet*, 2017). The increase in overweight- and obesity-related cancer gives new urgency to combating the already concerning growth in the global obesity rate.

Recent evidence strongly suggests that the gut microbiota is involved in the control of body weight, energy homeostasis and inflammation and thus plays an important role in the pathophysiology of obesity and obesity-related diseases (Turnbaugh *et al.*, 2006; Ding *et al.*, 2010). The human life cycle starts off with the microbial colonization of the gastrointestinal tract of the fetus and neonate. A 'healthy' gut microbiota (a state referred to as 'eubiosis') comprises a remarkable consortium of microbial communities whose profiles are blueprinted during the first six months of life (Palmer *et al.*, 2007). The healthy adult intestine harbours ~10^{14} microorganisms, a number almost 10 times that of the human cells present in a healthy body (Gill *et al.*, 2006), and the gut microbiome generally comprises ~100-fold more genes than the entire human genome (Tsai *et al.*, 2009). Infancy is a critical phase of colonization of the gut microbiota, a step that shapes the development of the early adult intestinal microbiota, potentially influencing host health in later years and perhaps throughout life (Kerr *et al.*, 2015). It is well acknowledged that the foundation of one's characteristic gut microbiota is laid down during infancy, immediately after birth. Many bacteria that characterize the adult microbiota, in fact, may have inhabited the gut already during the developmental window of early infancy; any perturbation in the establishment (improper colonization) and homeostasis results in a state referred to as 'dysbiosis', a condition in which one or more potentially harmful bacterial species may dominate. Dysbiosis in early life may affect the host's predisposition to various disorders, including obesity and obesity-related diseases (Putignani *et al.*, 2014). Thus, a robust microbiota is vital for sustaining optimal health, and vigorous efforts are being made to identify the factors that potentially affect the acquisition, establishment and maintenance of the microbiome throughout the host lifespan.

Intestinal microbiota composition in infants and children

More than 1000 bacterial species have been identified in the adult gastrointestinal tract (gut), with three phyla predominating: Bacteroides, Actinobacteria and Firmicutes. The colonization of the fetal gut begins *in utero* with swallowing of amniotic fluid. At that point, infants begin a lifelong relationship with their gut microbiota, as demonstrated by the detection of live bacteria in the meconium (Tsuji *et al.*, 2012). The initial exposure of the fetus to bacterial molecular patterns *in utero* serves to prime the immune system and/or the epithelium, facilitating appropriate postnatal responses to pathogens and commensals. Therefore, the impact of maternal microbiota may be more important than we have previously believed.

Immediately after birth, bacterial colonization of the gut proceeds rapidly; within one to two weeks of birth, the infant gut is populated with facultative aerobes such as enterobacteria, enterococci (mainly *Escherichia coli*, *Enterococcus faecalis*, *Enterococcus faecium*, *Klebsiella*, *Aeromonas* and *Acinetobacter*) and staphylococci. Colonization continues through breastfeeding until weaning, with additional bacteria accumulating in a stepwise manner with the appearance of obligate anaerobes (predominantly *Bifidobacterium*, *Bacteroides* and *Clostridium*) such that, by 2–3 years of age, the microbiota resembles the extremely dense and complex microbial colonization of an adult (Tsuji *et al.*, 2012). The infant microbiota is more variable in its composition and less stable over time than that of the adult (Mackie *et al.*, 1999; Sherman *et al.*, 2009). Our data show that an 'adult-type' intestinal microbiota profile may be established by four to five years of age (Wang *et al.*, 2015). Fig. 2.2 shows the main obligate anaerobes and facultative anaerobes that are typically detected in individuals from newborns to elders. *Bifidobacterium* remains a predominant component until the teenage years; although its presence subsequently gradually decreases through 80 years of age, *Bifidobacterium* remains a predominant bacterium despite ageing (Nomoto *et al.*, 2015).

Dysbiosis in infants and children

Infant microbial colonization can be significantly affected by exogenous factors including the mode of delivery [e.g. caesarean (C-) section vs. vaginal delivery] and feeding type (e.g. breastfeeding vs. formula feeding), antibiotic use and introduction of solid food. In addition, endogenous factors such as gestational age (e.g. full-term vs. premature delivery) and maternal nutrition (maternal undernutrition vs. obesity) can also influence the offspring's gut colonization. As a result, these factors may produce improper colonization; this effect, referred to as 'dysbiosis' (Penders *et al.*, 2006), can result in delays in colonization, increased susceptibility of newborns and infants to infections and altered immune homeostasis leading to the expression of immune-mediated disease (Gareau *et al.*, 2010) and metabolic diseases later in life (Walker *et al.*, 2013). In children, a high-fat (HF) diet is also known to change the composition of the colonic microflora, leading to increased body weight (Manco *et al.*, 2010).

Probiotics: definition from the FAO/WHO guideline

Between 2001 and 2002, two joint FAO and WHO documents were prepared with the aim of providing guidelines for assessing the safety and efficacy of 'probiotics', the term generally used to describe bacteria that are intended to provide beneficial effects when administered in humans (FAO/WHO, 2006). An analysis of data from a major bibliographic database (Web of Science) clearly shows that the use of the term 'probiotic(s)' as a keyword has shown a constant increase over the last 25 years. The term, a blend of Latin (pro = for, in favour of) and Greek (bios = life), has been defined as 'a live microbial feed supplement which beneficially affects the host animal by improving its intestinal balance' (Fuller, 1989).

Probiotics in the prevention and treatment of disease in infants and children

Pre-term low-birthweight (premature) infants

Pre-term birth is often the result of maternal microbial dysbiosis or infection (Goldenberg *et al.*, 2008); as a result, gut microbiota of premature infants differs from that of term infants (LaTuga *et al.*, 2011; Arboleya *et al.*, 2012). Gut colonization in pre-term infants has been described as delayed and aberrant and is generally characterized by lower diversity, greater fluctuation, a higher representation of potential pathogenic microorganisms typically encountered in hospital environments and a reduced representation of health-related commensal microorganisms (LaTuga *et al.*, 2011; Arboleya *et al.*, 2012). Pre-term neonates are frequently born by C-section, are minimally exposed to maternal microbes, receive frequent antibiotic treatment and delayed enteral feeding and are exposed to hospital environments for long periods of time (LaTuga *et al.*, 2011; Arboleya *et al.*, 2012). Given these factors and their immaturity, neonates are at increased risk of experiencing severe health condition, such as necrotizing enterocolitis (NEC) or sepsis, which can be fatal. In pre-term infants, gut microbiota shifts towards an increased presence of potential pathogens in the three weeks preceding an NEC or sepsis diagnosis; members of the *Corynebacterium*, *Klebsiella*, *Clostridium* or *Enterobacter* genera are among the most frequently identified microorganisms in these patients (Madan *et al.*, 2012; Mai *et al.*, 2013; Morrow *et al.*, 2013).

The early phase of life (during the first days after birth) may represent a unique window of opportunity for modulation of the microbiota by probiotics; such treatments may help reduce NEC and other morbidities and mortalities associated with prematurity.

Premature infants and probiotics

Several laboratories (e.g. Wang *et al.*, 2012), including our own (Yamashiro *et al.*, 2010), have reported that probiotic treatment can significantly reduce the risk of NEC and all-cause mortality without significant adverse effects.

Our study investigated the protective effects of *Bifidobacterium breve* (strain M16V) as a probiotic against NEC and infection in premature infants. Specifically, we performed a clinical study in 338 infants, consisting of 220 extremely low-birthweight (ELBW) and 118 very low-birthweight (VLBW) children, over a 5-year period. These patients were supplemented with *B. breve* starting several hours after birth (Bifido group). Two hundred and twenty-six premature infants served as controls. Infants of the Bifido group were administered *B. breve* as a daily dose of 1×10^9 cells/day. The incidence of NEC was significantly reduced in the Bifido group (nil) compared with that in control group (six cases, $P < 0.01$). The frequency of infection was also decreased significantly. Thus, administration of *B. breve* as a probiotic looks to be a very effective treatment for preventing NEC and infection in pre-term infants.

Dysbiosis in infants born by C-section

During birth and immediately after delivery, microorganisms from the mother and the surrounding environment inhabit the newborn's gut and initiate the development and establishment of complex microbiota. Infants born by C-section have been observed to acquire a different microbiota profile from infants born by vaginal delivery (Dominguez-Bello *et al.*, 2010) and the gut of vaginally born infants is known to be enriched in *Lactobacillus* spp. and *Bifidobacterium*, consistent with the profile of the maternal microbiota (Bäckhed *et al.*, 2015). In contrast, the gut of infants delivered by C-section lack this inoculum and is instead colonized by common skin and environmental microbes such as staphylococci and streptococci. In addition, we found that C-section-born infants had higher carriage of a-toxigenic and enterotoxigenic *Clostridium perfringens* (Nagpal *et al.*, 2017). Such dissimilarities in the gut microbe composition could still be detected at 19 years of age (Nagpal and Yamashiro, 2018). This dysbiosis is the prelude to a wide range of diseases, such as atopy and allergic diseases, obesity and T2DM, all of which are conditions that increasingly have been detected at higher frequencies in children (Putignani *et al.*, 2014).

Probiotic supplementation for C-section-induced dysbiosis

As mentioned above, the microbial composition of infants delivered by C-section differs from those born vaginally. One of the more important strategies for positively shaping the gut microbiota of these infants is microbial-based interventions. Systematic reviews (Anderson, 2015; Abdulkadir *et al.*, 2016) suggest that the administration of certain probiotics to premature infants reduces the incidence of NEC, yielding significant improvements in survival rates. However, reports of long-term effects of probiotics are sparse. One of the few such reports is a five-year follow-up study on probiotics for prevention of immunoglobulin E (IgE)-associated allergy in C-section-delivered children (Kuitunen *et al.*, 2009); these authors concluded that an allergy-preventative effect that extended to age five years was achieved with perinatal supplementation of probiotic bacteria to high-risk mothers and children born by C-section.

Children with obesity

Baker *et al.* (1993) first indicated the importance of fetal growth for development of metabolic syndrome in adulthood, with metabolic syndrome showing an increasing prevalence with decreasing birthweight. This study and others led to formulation of the developmental origins of health and diseases (DOHaD) hypothesis. Notably, both small-for-gestational-age (SGA) term-born infants and pre-term-born infants have an increased prevalence in childhood of several components of metabolic syndrome, including significantly higher glucose and triglyceride levels (Sato *et al.*, 2003; Hofman *et al.*, 2004; de Jong *et al.*, 2015) than term appropriate-for-gestational-age (AGA) children. Mechanisms underlying the association of microbiota with metabolic processes and obesity of the host organism are well known. Probiotic therapy has been proposed as a promising strategy for the management of metabolic disorders and obesity, given the ability of probiotics to restore microflora composition and ensure the maintenance of human health.

Childhood obesity has physical and psychological health consequences during childhood, can contribute to behavioural and emotional difficulties and reduces educational attainment (Lobstein *et al.*, 2006; Pizzi *et al.*, 2013). Importantly, obesity in childhood is a strong predictor of adult obesity with metabolic diseases including T2DM and cardiovascular disorders, etc. (Serino *et al.*, 2017). As such, obesity should be controlled in childhood to ensure a healthy adulthood. It is currently accepted that the gut microbiota plays a role in the pathophysiology of obesity, serving as an important environmental factor for generating energy and regulating inflammation (Ley *et al.*, 2005, 2006). Human studies have noted that, compared with non-obese adults, adults with obesity exhibit a lower ratio of Firmicutes and an elevated proportion of Bacteroidetes in the gut microbiota (Ley *et al.*, 2006). However, only limited data regarding microbial profiles are available for children with obesity (Karlsson *et al.*, 2012; Bervoets *et al.*, 2013). *Bifidobacterium* is known to be much more dominant than Bacteroidetes in the intestinal microbiota of children (Rahman *et al.*, 2012; Tsuji *et al.*, 2012) and reports in animal models have demonstrated that *Bifidobacterium* can suppress endotoxinaemia and thereby prevent type 1 diabetes (Cani *et al.*, 2007) while improving hypercholesterolaemia (Choi *et al.*, 2007). Given this information, we hypothesized that the elevation of the proportion of *Bifidobacterium* in the intestines of children could attenuate the frequency of such complications.

Probiotic supplementation in obese children

To test the above hypothesis, we conducted an open-label prospective clinical study to clarify the effects of *Lactobacillus casei* strain Shirota (LcS)-containing beverages in obese children (Nagata *et al.*, 2017). We compared the intestinal microbiota and organic acid levels in 12 obese [mean age 10.8 years; body mass index (BMI) *Z*-score (mean ± SD) 2.7 ± 1.7] and 22 control children [mean age 8.5 years; BMI *Z*-score (mean ± SD) 0.1 ± 0.7], including re-evaluation of the parameters after intervention in the obese children. The obese group underwent diet and exercise therapy for six months and then were given an LcS beverage daily for another 6 months; their body weights and serological markers were monitored. At baseline, the obese children (compared with the control group) exhibited significantly lower faecal concentrations of *Bifidobacterium* [obese group (mean ± SD) 7.9 ± 1.5 $\log_{10}$ cells/g vs. non-obese group (mean ± SD) 9.8 ± 0.5 $\log_{10}$ cells/g; $P < 0.01$) as well as the *Bacteroides fragilis* group, the *Atopobium* cluster and the *Lactobacillus gasseri* subgroup. Similarly, the obese children also exhibited significantly lower faecal concentrations of acetic acid at baseline [obese group (mean ± SD) 45.1 ± 16.9 µmol/g vs. non-obese group (mean SD) 57.9 ± 17.6 µmol/g; $P < 0.05$]. Compared with values at baseline, body weights were significantly decreased (mean ± SD −2.9 ± 4.6%; $P < 0.05$) and high-density lipoprotein (HDL) cholesterol levels were significantly increased (mean ± SD +11.1 ± 17.6%; $P < 0.05$) in the obese group after 6 months of ingestion of the LcS beverage. Furthermore, a significant increase in the faecal concentration of *Bifidobacterium* (mean ± SD 0.7 ± 1.2 $\log_{10}$ cells/g before ingestion vs. 9.1 ± 1.2 $\log_{10}$ cells/g after ingestion; $P < 0.01$) and an increase in the faecal acetic acid concentration (mean ± SD 7.0 ± 1.2 $\log_{10}$ cells/g before ingestion vs. 9.1 ± 1.2 $\log_{10}$ cells/g after ingestion; $P < 0.01$) were observed in the obese group after six months of ingestion of the LcS beverage. Thus, LcS consumption was associated with weight loss and improved lipid metabolism in obese children, a pattern that correlated with significant increases in the faecal *Bifidobacterium* numbers and acetic acid concentrations.

Children with chemotherapy-induced mucositis and probiotics

Mucositis

Mucositis, also referred to as mucosal barrier injury, is one of the most debilitating side effects of chemotherapy and radiotherapy (Bellm *et al.*, 2000). The condition is characterized by both inflammation and cell loss in the epithelial barrier lining the gastrointestinal tract (Sonis, 2004). Clinically, mucositis presenting with severe stomatitis and ulcer is associated with severe pain, appetite loss, diarrhoea, high fever/bacteraemia and malnutrition. These complications inevitably require total parenteral nutrition and intravenous infusion of broad-spectrum antibiotics, leading to longer hospitalization. Moreover, mucositis is a frequent reason for reducing the dosage of chemotherapeutics or postponing chemotherapy treatment, ultimately leading to a high mortality rate in cancer patients. Furthermore, chemotherapeutics has a detrimental effect on the intestinal microbial composition, drastically reducing the numbers of anaerobic bacteria (Van Vliet *et al.*, 2009).

Both inflammation and apoptosis of the mucosal barrier result in discontinuity of the barrier, thereby promoting bacterial translocation (Sonis, 2004). These findings strongly suggest that the commensal intestinal microbiota might play a pivotal role in chemotherapy-induced mucositis. Interventions are expected to restore eubiosis, and presumably would attenuate intestinal mucositis.

Considering the complexities of the mechanisms that underlie its aetiology, it seems likely that probiotic supplementation would be an effective therapy for mucositis.

Probiotic therapy in children with mucositis

To address this hypothesis, we performed a clinical study to ascertain whether administration of *B. breve* (strain BBG-01) attenuated intestinal mucositis in children with cancer being treated by chemotherapy (Yamashiro *et al.*, 2010). A placebo-controlled trial was performed in patients with malignancies admitted for chemotherapy ($n=42$), who were randomized into two groups receiving probiotics or placebo. The frequency of fever and the use of intravenous antibiotics were significantly lower in the BBG-01 group than in the placebo group. *B. breve* administration also enhanced the faecal concentration of anaerobes. Disruption of the intestinal microbiota after chemotherapy, as evidenced by an increase in the population levels of Enterobacteriaceae, was more pronounced in the placebo group than in the probiotic group. In conclusion, these data suggested that administration of *B. breve* is an effective approach to attenuating chemotherapy-induced mucositis in children with cancer.

Applicability of probiotics for healthy children

According to the 2001 FAO/WHO definition (FAO/WHO, 2006), probiotics are 'live microorganisms which when consumed in adequate amounts as part of food confer a health benefit on the host'. The evolution of knowledge in the past 20 years has demonstrated that a 'healthy' intestinal microbiota facilitates the development of well-balanced nutrition, metabolism, and immune tolerance in infants and children. Interventional studies on paediatric subjects with dysbiosis suggest that probiotics could be protective against the development of many diseases, as mentioned in the previous sections. However, reports on the effects of probiotic supplementation on the microbiota in healthy children are sparse. Therefore, we sought to investigate the effects of probiotic supplementation on the microbiota in healthy children. Specifically, we investigated the effects of daily intake of probiotic LcS-fermented milk on the faecal microbiota and intestinal environment of healthy preschool and school-age Japanese children before, during, and after six months of probiotic ingestion, and again six months later (i.e. six months following the termination of ingestion; the post-ingestion period) (Wang *et al.*, 2015).

In practice, the study consisted of an open-label trial in 23 children (14 boys, 9 girls; age (mean ± SD) 7.7 ± 2.4 years; BMI 19.6 ± 4.6 kg/m^2). We found that the composition of the intestinal microbiota of the healthy pre-school and school-age children resembled that of adults. During probiotic supplementation, the population levels of *Bifidobacterium* and total *Lactobacillus* increased significantly, whereas those of Enterobacteriaceae, *Staphylococcus*, and *C. perfringens* decreased significantly. Ingestion of LcS-fermented milk also yielded a significant increase in the faecal concentrations of organic acids and a significant decrease in faecal pH. However, the profiles of the faecal microbiota and intestinal environment were found to revert to the baseline (pre-ingestion) levels within 6 months following the cessation of probiotic intake. Notably, no adverse events due to LcS consumption were noted during the study interval. Therefore, we concluded that regular intake of an LcS-containing probiotic product may modify the gut microbiota composition and intestinal environment in pre-school and school-age healthy children while maintaining homeostasis of the microbiota.

To our knowledge, this trial was the first to demonstrate the beneficial effects of a probiotic supplementation in promoting a more favourable microbiota composition and intestinal environment in healthy pre-school and school-age children. These results warrant further investigations to evaluate the long-term effects of probiotic intake on the regulation of homeostasis of the intestinal microbiota and environment, and also to monitor the effects, if any, on the predisposition to several diseases in later years of life.

Conclusion

Perturbation of the gut microbiota composition (dysbiosis) in early life is known to be a pivotal factor for the subsequent development of obesity, obesity-related diseases such as T2DM, and allergic diseases. Probiotics may play important roles in maintaining gut homeostasis by modulating barrier function, immunity and metabolism and by beneficially modulating systemic biomarkers of various function. Nonetheless, many unanswered questions remain, because only a limited number of the clinical trials, most of small size, have been reported to date. However, the potential implications of this advancing field are promising.

References

Abdulkadir, B., Nelson, A., Skeath, T., Marrs, E.C., Perry, J.D., Cummings, S.P., Embleton, N.D., Berrington, J.E., and Stewart, C.J. (2016). Routine use of probiotics in preterm infants: Longitudinal impact on the microbiome and metabolome. Neonatology *109*, 239–247. https://doi.org/10.1159/000442936.

Anderson, S. (2015). Probiotics for preterm infants: A premature or overdue necrotizing enterocolitis prevention strategy? Neonatal Netw. *34*, 83–101. https://doi.org/10.1891/0730-0832.34.2.83.

Arboleya, S., Binetti, A., Salazar, N., Fernández, N., Solís, G., Hernández-Barranco, A., Margolles, A., de Los Reyes-Gavilán, C.G., and Gueimonde, M. (2012). Establishment and development of intestinal microbiota in preterm neonates. FEMS Microbiol. Ecol. *79*, 763–772. https://doi.org/10.1111/j.1574-6941.2011.01261.x.

Bäckhed, F., Roswall, J., Peng, Y., Feng, Q., Jia, H., Kovatcheva-Datchary, P., Li, Y., Xia, Y., Xie, H., Zhong, H., *et al.* (2015). Dynamics and stabilization of the human gut microbiome during the first year of life. Cell Host Microbe *17*, 690–703. https://doi.org/10.1016/j.chom.2015.04.004.

Barker, D.J., Hales, C.N., Fall, C.H., Osmond, C., Phipps, K., and Clark, P.M. (1993). Type 2 (non-insulin-dependent) diabetes mellitus, hypertension and hyperlipidaemia (syndrome X): relation to reduced fetal growth. Diabetologia *36*, 62–67.

Bellm, L.A., Epstein, J.B., Rose-Ped, A., Martin, P., and Fuchs, H.J. (2000). Patient reports of complications of bone marrow transplantation. Support. Care Cancer *8*, 33–39.

Bervoets, L., Van Hoorenbeeck, K., Kortleven, I., Van Noten, C., Hens, N., Vael, C., Goossens, H., Desager, K.N., and Vankerckhoven, V. (2013). Differences in gut microbiota composition between obese and lean children: A cross-sectional study. Gut Pathog. *5*, 10. https://doi.org/10.1186/1757-4749-5-10.

Cani, P.D., and Delzenne, N.M. (2007). Gut microflora as a target for energy and metabolic homeostasis. Curr. Opin. Clin. Nutr. Metab. Care *10*, 729–734. https://doi.org/10.1097/MCO.0b013e3282efdebb.

Choi, I., Kim, Y., Park, Y., Seog, H., and Choi, H. (2007). Anti-obesity activities of fermented soygerm isoflavones by Bifidobacterium breve. Biofactors *29*, 105–112.

de Jong, M., Cranendonk, A., and van Weissenbruch, M.M. (2015). Components of the metabolic syndrome in early childhood in very-low-birth-weight infants and term small and appropriate for gestational age infants. Pediatr. Res. *78*, 457–461. https://doi.org/10.1038/pr.2015.118.

Ding, S., Chi, M.M., Scull, B.P., Rigby, R., Schwerbrock, N.M., Magness, S., Jobin, C., and Lund, P.K. (2010). High-fat diet: Bacteria interactions promote intestinal inflammation which precedes and correlates with obesity and insulin resistance in mouse. PLOS ONE *5*, e12191. https://doi.org/10.1371/journal.pone.0012191.

Dominguez-Bello, M.G., Costello, E.K., Contreras, M., Magris, M., Hidalgo, G., Fierer, N., and Knight, R. (2010). Delivery mode shapes the acquisition and structure of the initial microbiota across multiple body habitats in newborns. Proc. Natl. Acad. Sci. U.S.A. *107*, 11971–11975. https://doi.org/10.1073/pnas.1002601107.

FAO/WHO (2006). Probiotics in food: Health and nutritional properties and guidelines for evaluation. Available online: http://www.fao.org/3/a-a0512e.pdf.

Fuller, R. (1989). Probiotics in man and animals. J. Appl. Bacteriol. *66*, 365–378.

Gareau, M.G., Sherman, P.M., and Walker, W.A. (2010). Probiotics and the gut microbiota in intestinal health and disease. Nat. Rev. Gastroenterol. Hepatol. *7*, 503–514. https://doi.org/10.1038/nrgastro.2010.117.

Gill, S.R., Pop, M., Deboy, R.T., Eckburg, P.B., Turnbaugh, P.J., Samuel, B.S., Gordon, J.I., Relman, D.A., Fraser-Liggett, C.M., and Nelson, K.E. (2006). Metagenomic analysis of the human distal gut microbiome. Science *312*, 1355–1359.

Goldenberg, R.L., Culhane, J.F., Iams, J.D., and Romero, R. (2008). Epidemiology and causes of preterm birth. Lancet *371*, 75–84. https://doi.org/10.1016/S0140-6736(08)60074-4.

Hofman, P.L., Regan, F., Jackson, W.E., Jefferies, C., Knight, D.B., Robinson, E.M., and Cutfield, W.S. (2004). Premature birth and later insulin resistance. N. Engl. J. Med. *351*, 2179–2186.

Karlsson, C.L., Onnerfält, J., Xu, J., Molin, G., Ahrné, S., and Thorngren-Jerneck, K. (2012). The microbiota of the gut in preschool children with normal and excessive body weight. Obesity *20*, 2257–2261. https://doi.org/10.1038/oby.2012.110.

Kerr, C.A., Grice, D.M., Tran, C.D., Bauer, D.C., Li, D., Hendry, P., and Hannan, G.N. (2015). Early life events influence whole-of-life metabolic health via gut microflora and gut permeability. Crit. Rev. Microbiol. *41*, 326–340. https://doi.org/10.3109/1040841X.2013.837863.

Kuitunen, M., Kukkonen, K., Juntunen-Backman, K., Korpela, R., Poussa, T., Tuure, T., Haahtela, T., and Savilahti, E. (2009). Probiotics prevent IgE-associated allergy until age 5 years in cesarean-delivered children but not in the total cohort. J. Allergy Clin. Immunol. *123*, 335–341. https://doi.org/10.1016/j.jaci.2008.11.019.

Lancet (2017). The link between cancer and obesity. Lancet. *390*, 1716. https://doi.org/10.1016/S0140-6736(17)32659-4.

LaTuga, M.S., Ellis, J.C., Cotton, C.M., Goldberg, R.N., Wynn, J.L., Jackson, R.B., and Seed, P.C. (2011). Beyond bacteria: A study of the enteric microbial consortium in extremely low birth weight infants. PLOS ONE *6*, e27858. https://doi.org/10.1371/journal.pone.0027858.

Ley, R.E., Bäckhed, F., Turnbaugh, P., Lozupone, C.A., Knight, R.D., and Gordon, J.I. (2005). Obesity alters gut microbial ecology. Proc. Natl. Acad. Sci. U.S.A. *102*, 11070–11075.

Ley, R.E., Turnbaugh, P.J., Klein, S., and Gordon, J.I. (2006). Microbial ecology: Human gut microbes associated with obesity. Nature *444*, 1022–1023.

Lobstein, T., and Jackson-Leach, R. (2006). Estimated burden of paediatric obesity and co-morbidities in Europe. Part 2. Numbers of children with indicators of obesity-related disease. Int. J. Pediatr. Obes. *1*, 33–41.

Mackie, R.I., Sghir, A., and Gaskins, H.R. (1999). Developmental microbial ecology of the neonatal gastrointestinal tract. Am. J. Clin. Nutr. *69*, 1035S–1045S. https://doi.org/10.1093/ajcn/69.5.1035s.

Madan, J.C., Salari, R.C., Saxena, D., Davidson, L., O'Toole, G.A., Moore, J.H., Sogin, M.L., Foster, J.A., Edwards, W.H., Palumbo, P., *et al.* (2012). Gut microbial colonisation in premature neonates predicts neonatal sepsis. Arch. Dis. Child. Fetal Neonatal Ed. *97*, F456–62. https://doi.org/10.1136/fetalneonatal-2011-301373.

Mai, V., Torrazza, R.M., Ukhanova, M., Wang, X., Sun, Y., Li, N., Shuster, J., Sharma, R., Hudak, M.L., and Neu, J. (2013). Distortions in development of intestinal microbiota associated with late onset sepsis in preterm infants. PLOS ONE *8*, e52876. https://doi.org/10.1371/journal.pone.0052876.

Manco, M., Putignani, L., and Bottazzo, G.F. (2010). Gut microbiota, lipopolysaccharides, and innate immunity in the pathogenesis of obesity and cardiovascular risk. Endocr. Rev. *31*, 817–844. https://doi.org/10.1210/er.2009-0030.

Morrow, A.L., Lagomarcino, A.J., Schibler, K.R., Taft, D.H., Yu, Z., Wang, B., Altaye, M., Wagner, M., Gevers, D., Ward, D.V., *et al.* (2013). Early microbial and metabolomic signatures predict later onset of necrotizing enterocolitis in preterm infants. Microbiome *1*, 13. https://doi.org/10.1186/2049-2618-1-13.

Nagata, S., Chiba, Y., Wang, C., and Yamashiro, Y. (2017). The effects of the *Lactobacillus casei* strain on obesity in children: A pilot study. Benef. Microbes *8*, 535–543. https://doi.org/10.3920/BM2016.0170.

Nagpal, R., Tsuji, H., Takahashi, T., Nomoto, K., Kawashima, K., Nagata, S., and Yamashiro, Y. (2017). Gut dysbiosis following C-section instigates higher colonisation of toxigenic Clostridium perfringens in infants. Benef. Microbes *8*, 353–365. https://doi.org/10.3920/BM2016.0216.

Nagpal, R., and Yamashiro, Y. (2018). Gut microbiota composition in healthy Japanese infants and young adults born by C-section. Ann. Nutr. Metab. 73, 4–11.

Nomoto, K., and Tsuji, H. (2015). [Establishment of a novel intestinal microbiota analysis system (YIF-SCAN®) and its application.] J. JSMUFF. *8*, 532–537.

Palmer, C., Bik, E.M., DiGiulio, D.B., Relman, D.A., and Brown, P.O. (2007). Development of the human infant intestinal microbiota. PLOS Biol. *5*, e177.

Pizzi, M.A., and Vroman, K. (2013). Childhood obesity: Effects on children's participation, mental health, and psychosocial development. Occup. Ther. Health Care *27*, 99–112. https://doi.org/10.3109/07380577.2013.784839.

Penders, J., Thijs, C., Vink, C., Stelma, F.F., Snijders, B., Kummeling, I., van den Brandt, P.A., and Stobberingh, E.E. (2006). Factors influencing the composition of the intestinal microbiota in early infancy. Pediatrics *118*, 511–521.

Putignani, L., Del Chierico, F., Petrucca, A., Vernocchi, P., and Dallapiccola, B. (2014). The human gut microbiota: A dynamic interplay with the host from birth to senescence settled during childhood. Pediatr. Res. *76*, 2–10. https://doi.org/10.1038/pr.2014.49.

Rahman, S., Nagata, S., Matsuda, K., Tsuji, H., Ogata, K., Ozaki, S., Kawashima, K., Nomoto, K., and Yamashiro, Y. (2012). Molecular biological studies of the origin of *Bifidobacterium* and *Lactobacillus* in neonatal feces. Int. J. Probiotics Prebiotics *7*, 91–98.

Sherman, P.M., Cabana, M., Gibson, G.R., Koletzko, B.V., Neu, J., Veereman-Wauters, G., Ziegler, E.E., and Walker, W.A. (2009). Potential roles and clinical utility of prebiotics in newborns, infants, and children: Proceedings from a global prebiotic summit meeting, New York City, June 27–28, 2008. J. Pediatr. *155*, S61–70. https://doi.org/10.1016/j.jpeds.2009.08.022.

Sonis, S.T. (2004). The pathobiology of mucositis. Nat. Rev. Cancer *4*, 277–284. https://doi.org/10.1038/nrc1318.

Soto, N., Bazaes, R.A., Peña, V., Salazar, T., Avila, A., Iñiguez, G., Ong, K.K., Dunger, D.B., and Mericq, M.V. (2003). Insulin sensitivity and secretion are related to catch-up growth in small-for-gestational-age infants at age 1 year: Results from a prospective cohort. J. Clin. Endocrinol. Metab. *88*, 3645–3650. https://doi.org/10.1210/jc.2002-030031.

Serino, M., Nicolas, S., Trabelsi, M.S., Burcelin, R., and Blasco-Baque, V. (2017). Young microbes for adult obesity. Pediatr. Obes. *12*, e28–e32. https://doi.org/10.1111/ijpo.12146.

Tsai, F., and Coyle, W.J. (2009). The microbiome and obesity: Is obesity linked to our gut flora? Curr. Gastroenterol. Rep. *11*, 307–313.

Tsuji, H., Oozeer, R., Matsuda, K., Matsuki, T., Ohta, T., Nomoto, K., Tanaka, R., Kawashima, M., Kawashima, K., Nagata, S., *et al.* (2012). Molecular monitoring of the development of intestinal microbiota in Japanese infants. Benef. Microbes 3, 113–125. https://doi.org/10.3920/BM2011.0038.

Turnbaugh, P.J., Ley, R.E., Mahowald, M.A., Magrini, V., Mardis, E.R., and Gordon, J.I. (2006). An obesity-associated gut microbiome with increased capacity for energy harvest. Nature *444*, 1027–1031.

van Vliet, M.J., Tissing, W.J., Dun, C.A., Meessen, N.E., Kamps, W.A., de Bont, E.S., and Harmsen, H.J. (2009). Chemotherapy treatment in pediatric patients with acute myeloid leukemia receiving antimicrobial prophylaxis leads to a relative increase of colonization with potentially pathogenic bacteria in the gut. Clin. Infect. Dis. *49*, 262–270. https://doi.org/10.1086/599346.

Wang, C., Nagata, S., Asahara, T., Yuki, N., Matsuda, K., Tsuji, H., Takahashi, T., Nomoto, K., and Yamashiro, Y. (2015). Intestinal microbiota profiles of healthy pre-school and school-age children and effects of probiotic supplementation. Ann. Nutr. Metab. *67*, 257–266. https://doi.org/10.1159/000441066.

Wang, Q., Dong, J., and Zhu, Y. (2012). Probiotic supplement reduces risk of necrotizing enterocolitis and mortality in preterm very low-birth-weight infants: An updated meta-analysis of 20 randomized, controlled trials. J. Pediatr. Surg. *47*, 241–248. https://doi.org/10.1016/j.jpedsurg.2011.09.064.

Walker, A.W., and Parkhill, J. (2013). Microbiology. Fighting obesity with bacteria. Science *341*, 1069–1070. https://doi.org/10.1126/science.1243787.

Yamashiro, Y., and Nagata, S. (2010). Beneficial microbes for premature infants, and children with malignancy undergoing chemotherapy. Benef. Microbes *1*, 357–365. https://doi.org/10.3920/BM2010.0035

Metaproteomics Approach to Gut Microbiota in Health and Disease

4

Bernardo Petriz[1,2*], Filipe Moura[2] and Octávio Franco[2,3]

[1]Research Group in Molecular and Physiologic Adaptations to Exercise, University Center of the Federal District, Brasília, Brazil.
[2]Centre for Proteomic and Biochemical Analysis, Graduate Studies in Biotechnology and Genomic Sciences, Catholic University of Brasília, Brasília, Brazil.
[3]S-Inova Biotech, Catholic University of Dom Bosco, Campo Grande, Brazil.

*Correspondence: bernardopetriz@gmail.com

https://doi.org/10.21775/9781910190937.04

Abstract

Phylotype profiling is conducted in a variety of biomedical research areas with significant impact on human health. So far, metagenomics has provided meaningful data linking the microbiota profile to the host health. However, profiling of the gut microbiome is limited to the identification of its constituents and their diversity. However, the complex molecular interactions between the microbiome and the gut environment affect the host metabolism, immunity and homeostasis. These interactions are being investigated by metaproteomics and metabolomics, as analytical approaches in the identification of large-scale proteins and metabolites, providing informative details and functional aspects of the microbiota and the gut environment. As a complementary approach to metagenomic and metatranscriptomic data, the metaproteomic analysis may also provide some new insights into the influence of the gut microbiota over the host biology. Considering the significant clinical potential of this field, the metaproteomic studies in the gut microbiota are still unexplored. In this chapter, we will present the main aspects and limitations of metaproteomics, as well as the perspectives in the analysis of the gut microbiome proteins in health and disease.

Introduction

The human gut microbiota is composed of trillions of microorganisms, most of which are represented by bacteria, archaea, non-pathological viruses and eukaryotes (Savage, 1977). This collection of microorganisms interacts symbiotically, in a molecular level, with the host, maintaining its physiological homeostasis. Furthermore, the gut microbiota is essential to the ongoing process of immune maturation and protection against pathogen colonization. In addition, the microbe community plays a prominent role in the digestion process, degrading and fermenting non-digestible polysaccharides and also acting as a source of amino acids, nutrients, vitamins and short-chain fatty acids (SCFAs) (Morrison and Preston, 2016).

Because of these extensive actions, the contribution of the gut microbiota to the maintenance of the host health its well established. This process may occur through direct contact from the bacteria species, present in the mucus layer, with the host lumen cells, or indirectly, through bacterial metabolites (e.g. butyrate, propionate, acetate) that are produced during digestion. Considering carbohydrates as one of the primary sources of energy substrates for human and microbial cells, the bacterial cells are responsible for degrading and fermenting non-digestible carbohydrates in the colon, yielding energy for their growth. In addition to energy delivery to the microbiome community and the host itself, end products of fermentation are known to have a significant effect on the gut environment and the host health (Morrison and Preston, 2016).

The equilibrium of this system is quite delicate, with alterations in the gut microbiota easily promoted by external factors (De Filippo *et al.*, 2010). When these disorders are chronic, a negative impact on one's health is usually observed. In this way, the development of several pathological conditions such as obesity, intestinal inflammatory disease and cardiovascular disease is being linked to perturbations in the gut microbiota (Carding *et al.*, 2015). As an example, gastrointestinal inflammatory disease is marked by a diminished microbial population and functional diversity (Hansen *et al.*, 2010), which also observed in faecal content from obese phenotypes (Turnbaugh *et al.*, 2006).

Several factors are implicated in the alterations of the gut microbiota quantitatively and qualitatively. Its known that extrinsic and intrinsic factors, which include exposure to several environmental elements such as pathogens, toxins, drugs, diet and even exercise, are known to alter the composition of the gut microbiome. However, the extension and magnitude of the influence of these elements is still widely studied, especially concerning its outcomes in human health (Lozupone *et al.*, 2012) (Fig. 4.1).

Because of its impact on the host health, the gut microbiota has been investigated using different perspectives. Whereas some focus on energy harvesting during digestion, others are interested in the close relation of the gut microbiota with innate immune maturation. In one way or another, solid evidence leads to a consensus that several physiological disorders or even the host health have a common agent, our microbiome community. In this way, the following sections will be dedicated to the recent studies of metaproteomics of the gut microbiota in the process of health maintenance. Naturally, the current and very useful functional analysis is a natural evolution of the exciting data that metagenomics has given us in the past decade.

From metagenomics to metaproteomics

Metagenome and metaprofiling

The development of next-generation sequencing (NGS) technologies has had a significant impact on microbial community studies, which were initially performed by Sanger sequencing. In a general way, both impacted the early studies in microbial communities. NGS technologies have some advantages over NGS technologies, which include the parallel sequencing of millions of DNA samples with different lengths, yields, running time and cost

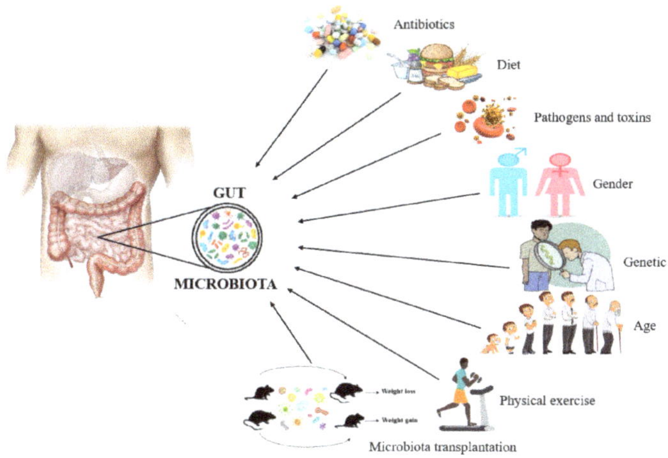

Figure 4.1 Main factors that influence the gut microbiota modulation. Although they are listed separately, these factors are in constant interaction, especially when there is a 'contact' between exogenous and endogenous factors (e.g. interaction of diet and gender or genetics of the individual).

(Escobar-Zepeda *et al.*, 2015). From pyrosequencing to reversible-termination sequencing, metagenomics was massively used by the microbial ecologist to explore diversity studies in a variety of research areas leading to a significant advance in metadata research.

To date (November 2017), 375 articles were found in PubMed under the research term 'metagenome' and 778 using the term 'metagenomics'. The popularization of metagenomic studies may be observed by the number of studies published in the last few years, covering several areas, from samples obtained in extreme environments such as plant-free hot and cold deserts, to the human environment as the gut microbiota, the main topic in this chapter (Fierer *et al.*, 2012; Wang *et al.*, 2015). Interestingly, samples collected in these extreme environments display lower levels of microbial phylogenetic and taxonomic diversity, sharing a similar characteristic observed in the gut from inflammatory bowel disease phenotypes and during dysbiosis (Carding *et al.*, 2015).

Next-generation sequencing in microbial ecology has permitted a series of important questions such as 'Who are they?' and 'What's going on there? to be answered. Therefore, in a general way, metagenomics has been successfully used to identify and establish microbial interactions in a variety of samples. These questions may be answered by amplicon sequencing and or by shotgun metagenomics. In prokaryotic samples, the first method is based on the amplification of specific regions of DNA from bacterial communities by specific primers such as the 16S component of the ribosomal RNA gene (16S rRNA). Besides, shotgun metagenomics is typically used to reconstruct fragments and or entire genomes from non-isolated organisms in a community. This approach is performed to characterize a significantly large number of coding and non-coding sequences. However, as evidenced by Escobar-Zepeda *et al.* (2015), the term 'metagenomics' may not fit well to amplicon sequence analysis. This is based on only one gene, instead of a collection of all genes present in the entire genomes from all organisms in a sample such as the whole bacterial community in the gut. In this sense, for taxonomy and phylogenetic purposes, the term 'meta-profiling' may be more adequate than metagenomics, once it refers to all members in a microbial community that are identified based on one gene or marker such as the 16S rRNA gene.

Independently of the platform used to obtain metadata, all of them present its strong points and its pitfalls, where the error rate from each sequencing method may considerably affect the characterization of the microbial community (Luo *et al.*, 2012). Furthermore, the sequencing process is commonly followed by filtering of low-quality reads to improve the data related to taxonomical classification and diversity calculations (Bokulich *et al.*, 2013).

Metaprofiling based on the 16S rRNA amplicon has been widely used to assess taxonomic and phylogenetic classifications from multiple replicates and samples obtained from longitudinal time points, which is very useful when focused on the gut microbiota (Escobar-Zepeda *et al.*, 2015). This type of phylogenetic marker has been highly used, which resulted in a significant representation of this marker in several databases such as the Ribosomal Database Project, Greengenes and Silva, permitting taxonomic annotation from diverse metagenomic samples. However, some sequencing errors are common to amplicon-dependent techniques such as primers bias, OTU assignment errors and some discrepancy when using different ribosomal variable regions. Besides, it is expected that the use of phylogenetic markers presents some pitfalls such as low resolution at the species level, the horizontal transfer of 16S rRNA genes and limited amplification from very low abundance genomes (Poretsky *et al.*, 2014).

Despite these pitfalls, meta-profiling present some significant advantages. With some limitations, it is possible to identify 'who they are' and the magnitude of the 'trouble'. In this sense, meta-profiling at different time points permits the establishment of reasonable relations between the profiled microbiota and the intervention that the host was submitted to. This type of study may give some information about the kinds of bacteria that are stimulated or not under certain external conditions such as dietary supplementation, pharmacological use or even different exercise regimens (Daniel *et al.*, 2014; Bermon *et al.*, 2015). Furthermore, the time-course analysis may also give some key information on the process of restoration of the gut microbiota by a qualitative and quantitative holistic, even considering the components that may contribute to its re-establishment such as probiotic supplementation (Delzenne *et al.*, 2011). Besides, meta-profiling may achieve its best result when it is linked to functional data and, when possible, with clinic observations, which may be used to determine some interventions such as:

- Dysbiosis followed by a prolonged period of antibiotic use: antibiotics are well known to deplete the gut microbiota, affecting not only the targeted pathogens. Here, perturbations in bacterial diversity, stability and resilience occur in response to an intense treatment with antibiotics, which is associated with periods of diarrhoea (Young and Schmidt, 2004). Clinical interventions may include probiotic supplementations and dietary control to re-establish the core microbiome.
- Alteration in gut microbiota after antibiotics: It has also been shown that the gut microbiota may be restored naturally, with the taxonomic composition of the gut microbiota restored after a few months after antibiotic treatment (e.g. four months after ciprofloxacin treatment); however, the same study indicated that several taxa failed to recover after more extended periods (e.g. six months after treatment) (Dethlefsen *et al.*, 2008). As in the first case, clinical intervention may include probiotic supplementation and dietary control.

- Dysbiosis and severe bowel inflammatory disease (IBD): IBDs are commonly accompanied by severe dysbiosis, where patients with IBD display a different microbiota (e.g. ↑ *P. aeruginosa*) from healthy patients (Shukla *et al.*, 2015). Clinical intervention may include microbiota transplantation and supplementation as mentioned before. Recent data have indicated that donors with *Bifidobacterium*-rich faecal content led to effective symbiosis in IBS patients after faecal transplantation (Mizuno *et al.*, 2017).
- Recurrent *Clostridium difficile* infection: recurrent infection with *C. difficile* presents a high failure rate of treatment with antibiotic therapy. As an alternative clinical intervention, infusion of donor faeces has been shown to be more effective than the use of vancomycin (van Nood *et al.*, 2013).
- Malnutrition and childhood undernutrition: a microbiota-directed therapy is being highly discussed and investigated to establish the feasible way to combat malnutrition and especially childhood undernutrition (Blanton *et al.*, 2016).

Metaproteomics

As described in the previous section, meta-profiling and metagenomics have been used to address a series of questions in the study of microbial consortia and communities from a variety of samples and experimental conditions. More recently, metaproteomics and metabolomics completed the 'omics' approach in the field of microbial ecology, raising some new questions (Xiong *et al.*, 2015a). Together, these novel strategies propose an in-depth, and high-throughput analysis that may uncover the metabolic activities of the certain microbial community *in situ* and under a variety of conditions. In this section, we will focus mainly on metaproteomics and its application in health and disease studies.

Together with genomics, transcriptomics and metabolomics, the proteomics techniques complete the main experimental approaches in the field of systems biology. In addition, the omics technologies such as metagenomics, metatranscriptomics and, more recently, metaproteomics and metabolomics are widely applied to microbial community studies (Xiong *et al.*, 2015a).

The application of proteomics in the study of the microbial community as initially described as community proteogenomics is the analysis of all proteins expressed within a specific consortium of bacteria and is used to resolve the functional aspects of microbial function. In this sense, more than 10 years ago, metaproteomics was defined as 'the large-scale characterization of the entire protein complements of environmental microbiota at a given point of time' (Wilmes and Bond, 2004). This definition remains pertinent to date.

Since its first use, by the identification of three proteins using traditional two-dimensional polyacrylamide gel electrophoresis (2D-PAGE) followed by *de novo* peptide sequencing (Ram *et al.*, 2005) compared with more recent shotgun liquid chromatography tandem mass spectrometry (LC-MS/MS) analysis, significant progress has been made, mainly in the analytical capability of MS-based proteomics. In addition, bioinformatic tools have been highly improved in the past decades (Wilmes *et al.*, 2015). It may be said that these improvements are permitting rapid progress in large-scale metaproteomics analysis from complex microbial samples such as soil, sludge and the human environment (Püttker *et al.*, 2015; Xiong *et al.*, 2015a; Zampieri *et al.*, 2016). Thus, when compared with the initial environmental microbial proteomic analysis, the actual accuracy and identification hits are considerably higher. However, the lack of well-known genome databases from these complex environmental samples is still a significant challenge to metaproteomics.

When considering the gastrointestinal environment, the complexity of metaproteomic analysis and its characterization is not different from that in other environments. It is estimated that more than 60,000,000 unique proteins are expressed in a faecal sample with ~21,000 taxa (Wilmes *et al.*, 2015). However, this environment itself presents its challenges to metaproteomics, being the gut locus known to promote severe proteolytic events, affecting not just the analytical process, but also leading to underestimated proteome identification (Lichtman *et al.*, 2013).

Almost a decade ago, 1340 non-redundant proteins were identified in the human faecal sample, which may be considered the first large-scale metaproteomic study on this type of sample. Interestingly, 30% of the observed spectra were linked to the human protein database, which indicated a significant contribution from the host proteome to the gut microbiota (Verberkmoes *et al.*, 2009). The combination of host and bacterial proteins is an additional challenge to any gut metaproteome study. In this sense, sample preparation and optimization before protein identification by MS is necessarily required.

Considering the main challenges to gut metaproteomics, some recommendations were proposed to improve protein separation, identification and quantification, which are well reviewed by Wilmes *et al.* (2015), Smirnov *et al.* (2016) and Petriz and Franco (2017). These recommendations are based on shotgun proteomics, being organized into three to four steps before MS analysis. Thus, the following must be considered:

1. sample depletion: depletion of host cells to enrich microbial cells avoiding cell-to-cell contamination (common to faecal samples)
2. protein extraction: here the focus is on extracting most proteins from the microbial community
3. cleaning sample: to remove any chemicals (e.g. detergents) used during the extraction process that may interfere and affect the operation of enzymatic digestion and mass spectrometry analysis
4. sample reduction: protein and peptide pre-fractionation before MS data acquirement and analysis.

As faecal samples contain a significant amount of Gram-positive and Gram-negative bacteria, as well as archaea, fungi, viruses and host cells, the depletion of bacterial species with more resistance to lysis methods should be avoided to preserve most of the other species. Moreover, it has been reported by Tanca *et al.* (2014) that bead-beating combined with freeze–thawing leads to enhanced protein extraction from yeast and Gram-negative bacteria.

After considering the sharp point and its pitfalls, it must also be understood that metaproteomics is still a novel analytical technique when applied to the microbial community. As previously stated, analytical limitations (e.g. host-to-bacterial cell interactions, unknown genome database) are added to diverse challenges involving the vast complexity of samples from the gastrointestinal tract (e.g. mucus, host tissue, faeces, high proteolytic locus). In this regard, to date, a large effort is being made to significantly improve all of the processes involved in sample preparation and throughout protein separation, identification and assignment by LC coupled to MS. A metaproteomic pipeline is presented in Fig. 4.2.

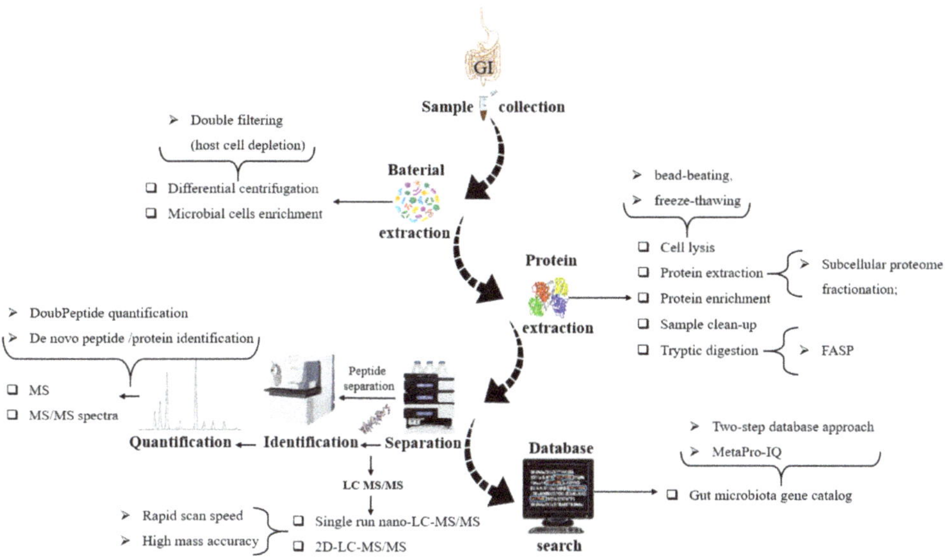

Figure 4.2 Typical workflow for metaproteomics analysis. After collection of the faecal sample, the process of protein extraction and non-interested components exclusion for analysis is started. Sequentially, similar to proteomics research, the extracted proteins are separated, identified, quantified and finally MS/MS search against genome databases.

Overcoming the main challenges in metaproteome characterization

The main limitations listed in the last section are quite challenging and time-consuming. Here, we list a series of technical steps that are being currently used to overcome them and improve metaproteomic data. It has been reported that dominant human peptides suppress ion signals from microbial peptides when both are co-eluted. This interference in MS/MS analysis misleads and underestimates the microbial proteomic data (Xiong *et al.*, 2015b).

Improvements over sample preparation and analytical pipeline

In faecal samples, a different centrifugation method is being used to enrich microbial cells (Verberkmoes *et al.*, 2009). A double filtering separation step is also being used when limited faecal sample is an issue. This process enriches the microbial biomass and depletes host cells and host proteins before proteomic analysis. This strategy was shown to improve low-abundant microbial proteins and preserve the relative distributions of protein abundance from each sample used in the experiment (Xiong *et al.*, 2015b).

Tanca *et al.* (2014) proposed an efficient analytical pipeline for metaproteomic characterization. Here, protein extraction was performed by combining bead-beating with freeze–thawing, followed by sample filtering for clean-up and enzymatic digestion. Finally, samples were submitted to a single run on LC coupled with MS (LC-MS) for peptide separation and identification. This pipeline was tested against a microbial mixture of bacteria and yeasts leading to 15,000 non-redundant peptide sequences per run. Furthermore, the same protocol was tested on mouse faecal samples leading to more than 13,000 non-redundant microbial peptides (false discovery rate of < 1%) from over 600 different microbial species.

This pipeline was shown to be less time-consuming when compared with state of the art metaproteomic techniques. As reviewed by Schneider and Riedel (2010), others have also suggested that a single run on a nano-LC-MS/MS is less time-consuming and laborious than the two-dimensional LC-MS/MS approach.

Improvements over database search engines

One of the significant challenges for shotgun metaproteomics is the absence of known databases for peptide spectra matching (PSM) using database search programs. In addition, strict filtering leads to more false negatives, reducing the number of peptide matches significantly. In this sense, to date, unknown gut microbiota genome database (Kolmeder *et al.*, 2012) is being used instead of high-sequence coverage databases (e.g. NCBI-nr). As reported by Jagtap *et al.* (2013) this step is less time-consuming, presenting an advantage over less sensitive peptide identification though false-discovery rate filtering (FDR). However, these synthetic databases are limited to a certain point because of the number of uncultivable intestinal microbes with unsequenced genomes.

On the other hand, a matched metagenome database may be used as an alternative to customized databases. In this sense, one successful option developed by Jagtap *et al.* (2013) combined target-only database to generate a smaller database to a target-decoy database research leading to significant improvement in peptide and protein identification. When compared with the one-step search method, this approach led to twice the number of high-confidence peptide sequence matches (PSMs) in the metaproteomic and proteogenomic analysis. These results were also superior independently of the database search program used for analysis.

Recently, Zhang *et al.* (2016) introduced a high-performance and universal workflow system for metaproteome identification and quantification termed as MetaPro-IQ. The human (9.9 million gut genes) or mouse (2.6 million gut genes) gut microbial gene catalogue was used as the database (almost completed and available at http://gigadb.org/). Authors found an average of 38% of the acquired MS spectra being identified in mice faecal samples, revealing 30,749 quantified protein groups. On the other hand, 30% of the acquires MS spectra were identified on the mucosal–lamina interface sample, leading to 19,011 protein groups for the human metaproteome. Furthermore, MetaPro-IQ was also tested for comparable identifications with the matched metagenome database search strategy. For this purpose, authors also compared the influence of a high-fat-diet on the gut microbiota, demonstrating that this pattern of diet leads to a distinct metaproteome (849 identified proteins) when compared with low-fat feeding.

Finally, a scalable set of databases (ComPIL) was also recently developed for metagenomic analysis. Here, three protein sequencing databases (ProtDB, MassDB and SeqDB) are integrated with a proteomic search engine (Brazmass) to perform fast matching of MS MudPIT spectra. This method was used to analyse the intracellular and secreted metaproteome from five human faecal samples over three technical replicates. More than 9000 proteins were identified per sample (Chatterjee *et al.*, 2016).

Furthermore, improvement in extraction and analytical pipelines together with the establishment of referenced gut microbial genome databases as the Human Microbiome Project (2012) or the Metagenomics of the Human Intestinal Tracts (MetaHIT) (Li *et al.*, 2014) are fundamental to the improvement of in-depth and ultra-metaproteomic profiling.

The gut metaproteome in health and disease

The gastrointestinal tract is by far the most complex biological micro-ecosystem in the human body. The role of the gut microbiota in this environment has been well described, mostly by the contribution of the several meta-profiling studies over the last decades. To this end, the identification of the host's microbial communities increased our knowledge of their effect on host health, as well as their contribution to pathological development (Nicholson *et al.*, 2012). Much was learned about dysbiosis and its influence on inflammatory bowel disease (Le Gall *et al.*, 2011), as well as the relationship between metabolic and cardiovascular diseases and the determined microbial and metabolic profile (Ryan *et al.*, 2017). Besides meta-profiling, metabolomics is also permitting a fast-track analysis of the diverse metabolites (e.g. SCFAs), with significant influence over food society and metabolic control, cognition, anxiety, depression and overall human health (Vernocchi *et al.*, 2016). All of these findings have indicated a straightforward crosstalk between the gut and the central nervous system, known today as the gut–brain axis (Foster and McVey Neufeld, 2013; Carabotti *et al.*, 2015).

It is secure to state that gut microbes are critical agents in host defence, immune maturation, regulation of inflammatory responses, nutrient processing and metabolic control and that their unbalanced state, known as dysbiosis, is associated with a large number of diseases (Carding *et al.*, 2015). However, meta-profiling is limited to identification, leading to a gap in functional interactions between the host genome and its microbe's genome. In this sense, environmental metaproteomics together with metabolomics may significantly complement meta-profiling. The identification of the set of proteins assigned to specific microbes, as well as tracing their influence on the host biology, may lead to an in-depth understanding of the host response to physiological and pathological insults. In sum, metaproteomics and metabolomics deliver a comprehensive 'functional' data that may lead us to possible clinical outcomes with regard to pathological prevention and treatment with a likely impact on human health (Haange and Jehmlich, 2016).

Inflammatory bowel disease

As unfunctional interactions between the host cells and the gut microbiota may contribute to the pathogenesis of much gastrointestinal illness, the first metaproteomic characterization on this locus was performed to characterize the metaproteome of human mucosa–luminal interface from several intestinal *loci* (Li *et al.*, 2011). Through quantitative shotgun proteomics, authors were able to identify a biogeographic profile associated with 49 proteins from the proximal and distal colon. In another study, the mucosa from healthy subjects was compared with inflamed mucosa inherent to subjects with IBD (Crohn's disease, ulcerative colitis). It was found that a large number of the identified bacteria (35%) differentiated the subjects by the disease type (Presley *et al.*, 2012). A significant variation in the intestinal epithelial barrier proteins was found between Crohn's disease patients and healthy individuals, with 2900 microbial proteins explicitly assigned to Crohn's disease when compared with healthy patients (Erickson *et al.*, 2012).

It must be noted that faecal samples are an optimal sample resource for metaproteome characterization for many reasons. As well as being an easily obtained specimen, it is also easily collected at several time points, permitting the investigation of the gut environment under diverse physiological and pathological conditions. In this sense, classical metagenomic profiling (Petriz *et al.*, 2014; Campbell *et al.*, 2016) would benefit from metaproteomic characterization.

Recently, an elegant study used a reverse-phase LC-MS/MS to characterize the proteomic profile of the luminal content from mouse stomach, jejunum, ileum, caecum and proximal colon under three different colonization states: germ-free, single commensal microbe (*Bacteroides thetaiotaomicron*) and conventionally raised pathogen-free microbiota. Authors have selected the *Bacteroides thetaiotaomicron* species based on its regular and abundant status in the human microbiota. The LC-MS/MS was able to identify 853 unique proteins being common across all three colonization states and the five locations. The proteases anionic trypsin-2 and chymotrypsin-like elastase family member 2A were the most abundant proteins in all locations and colonization states. Furthermore, this study has pointed that simpler microbiota (germ-free and moon colonization) presents a higher overlap on their protein profile; the variation in protein abundance occurs primarily by location rather than by its colonization state; and that proteins identified in faecal samples are originated throughout the gastrointestinal tract. In sum, this was the first study to present an overall proteomic characterization across the digestive tracts and under different colonization states. One possible outcome of studies like this is the possibility of identifying candidate proteins that may act as local biomarkers of the microbiota under pathological insult or a variety of host colonization states. Likewise, other described studies in this section, the complement of metaproteomic data with metagenome profiling (16S rRNA) permits the clustering of specific microbial taxa with host protein expression.

Future perspectives in metaproteomic research

Despite its great clinical potential, the use of metaproteomics in the gut environment is still far from achieving its full potential. However, the well-designed experiments may deliver valuable data with a significant impact on human health. It is noteworthy that metaproteomics and metabolomics are still struggling with technical challenges, especially when considering sample preparation and analytical data acquisition and quantification. However, as described in the previous sections, great progress has been made across all of the specific steps. It has also been suggested that, to achieve an in-depth characterization of several microbial communities, more technical improvements are required (Wilmes *et al.*, 2015). It is noteworthy that a recent ultra-deep metaproteomic analysis of human stool identified more than 20,500 protein groups per sample. This indicates that the continuous improvement of straightforward pipelines and analytical procedures will reduce the pre-existing barriers to the metaproteomic approach, leading it to being less time-consuming. In addition, the ultra-deep metaproteomic analysis may be a tendency in this field, because of the great number of data generated.

As occurred with the other meta-omics, rapid and significant technical advances are expected over the coming years. This may be observed by the reduction of sample processing and analysis, with possible integration with the other meta-omics, as it is currently being done with metabolomics by tracking several metabolites from diverse host biological sources (e.g. saliva, plasma, urine, facees) and their microbes and their co-metabolites (Smirnov *et al.*, 2016). On the other hand, integrating metabolomics with metaproteomics for the analysis of faecal samples will be of great significance to track metabolites, and co-metabolites originated from the host, their microbes and food degradation components. As mentioned earlier, these components support not only the gut–brain axis with a significant contribution to metabolic regulation and neuroendocrine homeostasis, but several other organs on health and disease, as well reviewed by Vernocchi *et al.* (2016).

Therefore, as currently observed in metagenome profiling, the combination of metaproteomics with metabolomics may give more information over the contribution of the gut microbiota to the pathogenesis of several diseases such as metabolic disorders and cardiovascular disease (Yamashita *et al.*, 2015; Sanchez-Alcoholado *et al.*, 2017; Tang *et al.*, 2017). However, most of the publications concerning faecal metagenome profiling, metaproteomics and metabolomics are associated with the context of pathological insults, such as mucosa inflammation (IBD), cancer, infections, obesity and more recently cardiovascular disease. As discussed in this chapter, most of this research is being performed using 16S rRNA profiling. Thus, the field of metaproteomics is less explored, indicating a vast area to be improved and explored.

References

Bermon, S., Petriz, B., Kajėnienė, A., Prestes, J., Castell, L., and Franco, O.L. (2015). The microbiota: An exercise immunology perspective. Exerc. Immunol. Rev. *21*, 70–79.

Blanton, L.V., Barratt, M.J., Charbonneau, M.R., Ahmed, T., and Gordon, J.I. (2016). Childhood undernutrition, the gut microbiota, and microbiota-directed therapeutics. Science *352*, 1533. https://doi.org/10.1126/science.aad9359.

Bokulich, N.A., Subramanian, S., Faith, J.J., Gevers, D., Gordon, J.I., Knight, R., Mills, D.A., and Caporaso, J.G. (2013). Quality-filtering vastly improves diversity estimates from Illumina amplicon sequencing. Nat. Methods *10*, 57–59. https://doi.org/10.1038/nmeth.2276.

Campbell, S.C., Wisniewski, P.J., Noji, M., McGuinness, L.R., Häggblom, M.M., Lightfoot, S.A., Joseph, L.B., and Kerkhof, L.J. (2016). The effect of diet and exercise on intestinal integrity and microbial diversity in mice. PLOS ONE *11*, e0150502. https://doi.org/10.1371/journal.pone.0150502.

Carabotti, M., Scirocco, A., Maselli, M.A., and Severi, C. (2015). The gut-brain axis: Interactions between enteric microbiota, central and enteric nervous systems. Ann. Gastroenterol. *28*, 203–209.

Carding, S., Verbeke, K., Vipond, D.T., Corfe, B.M., and Owen, L.J. (2015). Dysbiosis of the gut microbiota in disease. Microb. Ecol. Health Dis. *26*, 26191. https://doi.org/10.3402/mehd.v26.26191.

Chatterjee, S., Stupp, G.S., Park, S.K., Ducom, J.C., Yates, J.R., Su, A.I., and Wolan, D.W. (2016). A comprehensive and scalable database search system for metaproteomics. BMC Genomics *17*, 642. https://doi.org/10.1186/s12864-016-2855-3.

Daniel, H., Gholami, A.M., Berry, D., Desmarchelier, C., Hahne, H., Loh, G., Mondot, S., Lepage, P., Rothballer, M., Walker, A., *et al.* (2014). High-fat diet alters gut microbiota physiology in mice. ISME J. *8*, 295–308. https://doi.org/10.1038/ismej.2013.155.

De Filippo, C., Cavalieri, D., Di Paola, M., Ramazzotti, M., Poullet, J.B., Massart, S., Collini, S., Pieraccini, G., and Lionetti, P. (2010). Impact of diet in shaping gut microbiota revealed by a comparative study in children from Europe and rural Africa. Proc. Natl. Acad. Sci. U.S.A. *107*, 14691–14696. https://doi.org/10.1073/pnas.1005963107.

Delzenne, N.M., Neyrinck, A.M., Bäckhed, F., and Cani, P.D. (2011). Targeting gut microbiota in obesity: Effects of prebiotics and probiotics. Nat. Rev. Endocrinol. *7*, 639–646. https://doi.org/10.1038/nrendo.2011.126.

Dethlefsen, L., Huse, S., Sogin, M.L., and Relman, D.A. (2008). The pervasive effects of an antibiotic on the human gut microbiota, as revealed by deep 16S rRNA sequencing. PLOS Biol. *6*, e280. https://doi.org/10.1371/journal.pbio.0060280.

Erickson, A.R., Cantarel, B.L., Lamendella, R., Darzi, Y., Mongodin, E.F., Pan, C., Shah, M., Halfvarson, J., Tysk, C., Henrissat, B., *et al.* (2012). Integrated metagenomics/metaproteomics reveals human host-microbiota signatures of Crohn's disease. PLOS ONE *7*, e49138. https://doi.org/10.1371/journal.pone.0049138.

Escobar-Zepeda, A., Vera-Ponce de León, A., and Sanchez-Flores, A. (2015). The road to metagenomics: From microbiology to DNA sequencing technologies and bioinformatics. Front. Genet. *6*, 348. https://doi.org/10.3389/fgene.2015.00348.

Fierer, N., Leff, J.W., Adams, B.J., Nielsen, U.N., Bates, S.T., Lauber, C.L., Owens, S., Gilbert, J.A., Wall, D.H., and Caporaso, J.G. (2012). Cross-biome metagenomic analyses of soil microbial communities and their functional attributes. Proc. Natl. Acad. Sci. U.S.A. *109*, 21390–21395. https://doi.org/10.1073/pnas.1215210110.

Foster, J.A., and McVey Neufeld, K.A. (2013). Gut-brain axis: How the microbiome influences anxiety and depression. Trends Neurosci. *36*, 305–312. https://doi.org/10.1016/j.tins.2013.01.005.

Haange, S.B., and Jehmlich, N. (2016). Proteomic interrogation of the gut microbiota: Potential clinical impact. Expert Rev. Proteomics *13*, 535–537. https://doi.org/10.1080/14789450.2016.1190652.

Hansen, J., Gulati, A., and Sartor, R.B. (2010). The role of mucosal immunity and host genetics in defining intestinal commensal bacteria. Curr. Opin. Gastroenterol. *26*, 564–571. https://doi.org/10.1097/MOG.0b013e32833f1195.

Human Microbiome Project Consortium (2012). Structure, function and diversity of the healthy human microbiome. Nature *486*, 207–214. https://doi.org/10.1038/nature11234.

Jagtap, P., Goslinga, J., Kooren, J.A., McGowan, T., Wroblewski, M.S., Seymour, S.L., and Griffin, T.J. (2013). A two-step database search method improves sensitivity in peptide sequence matches for metaproteomics and proteogenomics studies. Proteomics *13*, 1352–1357. https://doi.org/10.1002/pmic.201200352.

Kolmeder, C.A., de Been, M., Nikkilä, J., Ritamo, I., Mättö, J., Valmu, L., Salojärvi, J., Palva, A., Salonen, A., and de Vos, W.M. (2012). Comparative metaproteomics and diversity analysis of human intestinal microbiota testifies for its temporal stability and expression of core functions. PLOS ONE *7*, e29913. https://doi.org/10.1371/journal.pone.0029913.

Le Gall, G., Noor, S.O., Ridgway, K., Scovell, L., Jamieson, C., Johnson, I.T., Colquhoun, I.J., Kemsley, E.K., and Narbad, A. (2011). Metabolomics of fecal extracts detects altered metabolic activity of gut microbiota in ulcerative colitis and irritable bowel syndrome. J. Proteome Res. *10*, 4208–4218. https://doi.org/10.1021/pr2003598.

Li, J., Jia, H., Cai, X., Zhong, H., Feng, Q., Sunagawa, S., Arumugam, M., Kultima, J.R., Prifti, E., Nielsen, T., *et al.* (2014). An integrated catalog of reference genes in the human gut microbiome. Nat. Biotechnol. *32*, 834–841. https://doi.org/10.1038/nbt.2942.

Li, X., LeBlanc, J., Truong, A., Vuthoori, R., Chen, S.S., Lustgarten, J.L., Roth, B., Allard, J., Ippoliti, A., Presley, L.L., *et al.* (2011). A metaproteomic approach to study human-microbial ecosystems at the mucosal luminal interface. PLOS ONE *6*, e26542. https://doi.org/10.1371/journal.pone.0026542.

Lichtman, J.S., Marcobal, A., Sonnenburg, J.L., and Elias, J.E. (2013). Host-centric proteomics of stool: A novel strategy focused on intestinal responses to the gut microbiota. Mol. Cell Proteomics *12*, 3310–3318. https://doi.org/10.1074/mcp.M113.029967.

Lozupone, C.A., Stombaugh, J.I., Gordon, J.I., Jansson, J.K., and Knight, R. (2012). Diversity, stability and resilience of the human gut microbiota. Nature *489*, 220–230. https://doi.org/10.1038/nature11550.

Luo, C., Tsementzi, D., Kyrpides, N., Read, T., and Konstantinidis, K.T. (2012). Direct comparisons of Illumina vs. Roche 454 sequencing technologies on the same microbial community DNA sample. PLOS ONE *7*, e30087. https://doi.org/10.1371/journal.pone.0030087.

Mizuno, S., Masaoka, T., Naganuma, M., Kishimoto, T., Kitazawa, M., Kurokawa, S., Nakashima, M., Takeshita, K., Suda, W., Mimura, M., *et al.* (2017). Bifidobacterium-rich fecal donor may be a positive predictor for successful fecal microbiota transplantation in patients with irritable bowel syndrome. Digestion *96*, 29–38. https://doi.org/10.1159/000471919.

Morrison, D.J., and Preston, T. (2016). Formation of short chain fatty acids by the gut microbiota and their impact on human metabolism. Gut Microbes *7*, 189–200. https://doi.org/10.1080/19490976.2015.1134082.

Nicholson, J.K., Holmes, E., Kinross, J., Burcelin, R., Gibson, G., Jia, W., and Pettersson, S. (2012). Host-gut microbiota metabolic interactions. Science *336*, 1262–1267. https://doi.org/10.1126/science.1223813.

Petriz, B.A., Castro, A.P., Almeida, J.A., Gomes, C.P., Fernandes, G.R., Kruger, R.H., Pereira, R.W., and Franco, O.L. (2014). Exercise induction of gut microbiota modifications in obese, non-obese and hypertensive rats. BMC Genomics *15*, 511. https://doi.org/10.1186/1471-2164-15-511.

Petriz, B.A., and Franco, O.L. (2017). Metaproteomics as a complementary approach to gut microbiota in health and disease. Front. Chem. *5*, 4. https://doi.org/10.3389/fchem.2017.00004.

Poretsky, R., Rodriguez-R, L.M., Luo, C., Tsementzi, D., and Konstantinidis, K.T. (2014). Strengths and limitations of 16S rRNA gene amplicon sequencing in revealing temporal microbial community dynamics. PLOS ONE *9*, e93827. https://doi.org/10.1371/journal.pone.0093827.

Presley, L.L., Ye, J., Li, X., Leblanc, J., Zhang, Z., Ruegger, P.M., Allard, J., McGovern, D., Ippoliti, A., Roth, B., *et al.* (2012). Host-microbe relationships in inflammatory bowel disease detected by bacterial and metaproteomic analysis of the mucosal-luminal interface. Inflamm. Bowel Dis. *18*, 409–417. https://doi.org/10.1002/ibd.21793.

Püttker, S., Kohrs, F., Benndorf, D., Heyer, R., Rapp, E., and Reichl, U. (2015). Metaproteomics of activated sludge from a wastewater treatment plant - a pilot study. Proteomics *15*, 3596–3601. https://doi.org/10.1002/pmic.201400559.

Ram, R.J., Verberkmoes, N.C., Thelen, M.P., Tyson, G.W., Baker, B.J., Blake, R.C., Shah, M., Hettich, R.L., and Banfield, J.F. (2005). Community proteomics of a natural microbial biofilm. Science *308*, 1915–1920.

Ryan, P.M., London, L.E., Bjorndahl, T.C., Mandal, R., Murphy, K., Fitzgerald, G.F., Shanahan, F., Ross, R.P., Wishart, D.S., Caplice, N.M., *et al.* (2017). Microbiome and metabolome modifying effects of several cardiovascular disease interventions in apo-E-/- mice. Microbiome *5*, 30. https://doi.org/10.1186/s40168-017-0246-x.

Sanchez-Alcoholado, L., Castellano-Castillo, D., Jordán-Martínez, L., Moreno-Indias, I., Cardila-Cruz, P., Elena, D., Muñoz-Garcia, A.J., Queipo-Ortuño, M.I., and Jimenez-Navarro, M. (2017). Role of gut microbiota on cardio-metabolic parameters and immunity in coronary artery disease patients with and without type-2 diabetes mellitus. Front. Microbiol. *8*, 1936. https://doi.org/10.3389/fmicb.2017.01936.

Savage, D.C. (1977). Microbial ecology of the gastrointestinal tract. Annu. Rev. Microbiol. *31*, 107–133. https://doi.org/10.1146/annurev.mi.31.100177.000543.

Schneider, T., and Riedel, K. (2010). Environmental proteomics: Analysis of structure and function of microbial communities. Proteomics *10*, 785–798. https://doi.org/10.1002/pmic.200900450.

Shukla, R., Ghoshal, U., Dhole, T.N., and Ghoshal, U.C. (2015). Fecal microbiota in patients with irritable bowel syndrome compared with healthy controls using real-time polymerase chain reaction: An evidence of dysbiosis. Dig. Dis. Sci. *60*, 2953–2962. https://doi.org/10.1007/s10620-015-3607-y.

Smirnov, K.S., Maier, T.V., Walker, A., Heinzmann, S.S., Forcisi, S., Martinez, I., Walter, J., and Schmitt-Kopplin, P. (2016). Challenges of metabolomics in human gut microbiota research. Int. J. Med. Microbiol. *306*, 266–279.

Tanca, A., Palomba, A., Pisanu, S., Deligios, M., Fraumene, C., Manghina, V., Pagnozzi, D., Addis, M.F., and Uzzau, S. (2014). A straightforward and efficient analytical pipeline for metaproteome characterization. Microbiome 2, 49. https://doi.org/10.1186/s40168-014-0049-2.

Tang, W.H., Kitai, T., and Hazen, S.L. (2017). Gut microbiota in cardiovascular health and disease. Circ. Res. *120*, 1183–1196. https://doi.org/10.1161/CIRCRESAHA.117.309715.

Turnbaugh, P.J., Ley, R.E., Mahowald, M.A., Magrini, V., Mardis, E.R., and Gordon, J.I. (2006). An obesity-associated gut microbiome with increased capacity for energy harvest. Nature *444*, 1027–1031.

van Nood, E., Vrieze, A., Nieuwdorp, M., Fuentes, S., Zoetendal, E.G., de Vos, W.M., Visser, C.E., Kuijper, E.J., Bartelsman, J.F., Tijssen, J.G., *et al.* (2013). Duodenal infusion of donor feces for recurrent *Clostridium difficile*. N. Engl. J. Med. *368*, 407–415. https://doi.org/10.1056/NEJMoa1205037.

Verberkmoes, N.C., Russell, A.L., Shah, M., Godzik, A., Rosenquist, M., Halfvarson, J., Lefsrud, M.G., Apajalahti, J., Tysk, C., Hettich, R.L., *et al.* (2009). Shotgun metaproteomics of the human distal gut microbiota. ISME J. *3*, 179–189. https://doi.org/10.1038/ismej.2008.108.

Vernocchi, P., Del Chierico, F., and Putignani, L. (2016). Gut microbiota profiling: Metabolomics based approach to unravel compounds affecting human health. Front. Microbiol. *7*, 1144. https://doi.org/10.3389/fmicb.2016.01144.

Wang, W.L., Xu, S.Y., Ren, Z.G., Tao, L., Jiang, J.W., and Zheng, S.S. (2015). Application of metagenomics in the human gut microbiome. World J. Gastroenterol. *21*, 803–814. https://doi.org/10.3748/wjg.v21.i3.803.

Wilmes, P., and Bond, P.L. (2004). The application of two-dimensional polyacrylamide gel electrophoresis and downstream analyses to a mixed community of prokaryotic microorganisms. Environ. Microbiol. *6*, 911–920. https://doi.org/10.1111/j.1462-2920.2004.00687.x.

Wilmes, P., Heintz-Buschart, A., and Bond, P.L. (2015). A decade of metaproteomics: Where we stand and what the future holds. Proteomics *15*, 3409–3417. https://doi.org/10.1002/pmic.201500183.

Xiong, W., Abraham, P.E., Li, Z., Pan, C., and Hettich, R.L. (2015a). Microbial metaproteomics for characterizing the range of metabolic functions and activities of human gut microbiota. Proteomics *15*, 3424–3438. https://doi.org/10.1002/pmic.201400571.

Xiong, W., Giannone, R.J., Morowitz, M.J., Banfield, J.F., and Hettich, R.L. (2015b). Development of an enhanced metaproteomic approach for deepening the microbiome characterization of the human infant gut. J. Proteome Res. *14*, 133–141. https://doi.org/10.1021/pr500936p.

Yamashita, T., Kasahara, K., Emoto, T., Matsumoto, T., Mizoguchi, T., Kitano, N., Sasaki, N., and Hirata, K. (2015). Intestinal immunity and gut microbiota as therapeutic targets for preventing atherosclerotic cardiovascular diseases. Circ. J. *79*, 1882–1890. https://doi.org/10.1253/circj.CJ-15-0526.

Young, V.B., and Schmidt, T.M. (2004). Antibiotic-associated diarrhea accompanied by large-scale alterations in the composition of the fecal microbiota. J. Clin. Microbiol. *42*, 1203–1206.

Zampieri, E., Chiapello, M., Daghino, S., Bonfante, P., and Mello, A. (2016). Soil metaproteomics reveals an inter-kingdom stress response to the presence of black truffles. Sci. Rep. *6*, 25773. https://doi.org/10.1038/srep25773.

Zhang, X., Ning, Z., Mayne, J., Moore, J.I., Li, J., Butcher, J., Deeke, S.A., Chen, R., Chiang, C.K., Wen, M., *et al.* (2016). MetaPro-IQ: A universal metaproteomic approach to studying human and mouse gut microbiota. Microbiome *4*, 31. https://doi.org/10.1186/s40168-016-0176-z

Gut Microbiome: The Role in Neurodevelopment

5

Surajit Pathak[1]*, Suhanya Veronica Prasad[1], Francesco Marotta[2]*, Woo-Sik Jeong[3], Ramachandran Murugesan[1] and Antara Banerjee[1]

[1]Faculty of Allied Health Sciences, Chettinad Academy of Research and Education, Kelambakkam, Chennai-603103, India.
[2]ReGenera R&D International for Ageing Intervention, Milano-Beijing and VCC Preventative Medical Promotion Foundation, Beijing, China.
[3]Department of Food and Life Sciences, Inje University, Gimhae, South Korea.

*Correspondence: surajit.pathak@gmail.com and fmarchimede@libero.it

https://doi.org/10.21775/9781910190937.05

Abstract

The microbiota–gut–brain axis is a dynamic relationship in the body that functions to maintain homeostasis. Increasing evidence implicates the significance of gut microbes in neurodevelopment and behaviour. These indigenous organisms communicate to the brain via multiple routes, which include neural and humoral pathways. Modulations of the commensal microbes have been found to have deleterious effects on the physiological functioning of the brain. Considerably, over the past decade investigations on the complexity and diversity of microorganisms that colonize the gut exhibited profound impacts on cognitive processes and social conduct. The Human Microbiome Project established in 2008 enables understanding of the comprehensive characteristics and role of microbes in the health and disease states of the brain. The microbial metabolites, cytokines, peptides and neurotransmitters produced in the gut enhance the brain response. The crosstalk between the host mucosal system and the intestinal microbiome has been studied in various neurological diseases. In this review we have highlighted the role of the microbiota in normal brain development and dysfunction leading to several diseases. Gaining a better understanding of this bidirectional pathway can provide an insight into innovative and therapeutic strategies for neurodevelopmental and behavioural disorders.

Introduction

Humans are believed to have a strong communication with the indigenous microbes found in the body. In time, they have co-evolved and appear to possess a complex genome that is about 100 times more in number than the human genome. The myriad range of microorganisms abundantly present in the body is known to perform various beneficial and essential functions both autonomously and co-operatively. Innumerable experimental studies have investigated the participation of microorganisms in host defence, regulation of immunity and their role in the occurrence of autoimmune diseases by comparing germ-free animals with those exposed to pathogenic bacterial infections. About 10–100 trillion microorganisms are known to inhabit the gastrointestinal (GI) tract. Interestingly the microbial organisms residing in the gut region are susceptible to change by the external environment. It has been a decade since research in gut microbes has exploded and it has certainly established the two-way line of connection linking the gut bacteria and brain. An altered gut microbiome along with perturbation of gut–brain axis has been implicated in the aetiology and pathophysiology of several neurodevelopmental and neuropsychiatric disorders. This chapter aims to provide convincing evidence for understanding the complex interaction between gut microbes and the brain; these minute organisms have the ability to regulate the fundamental development of the brain as well as functioning, behaviour and mood.

Prevalence of neurodevelopmental and neuropsychiatric disorders

According to epidemiologists from the Centers for Disease Control and Prevention (CDC), in collaboration with researchers from the Health Resources and Services Administration (HRSA), the number of people being identified with a developmental disability is shockingly high and it continues to increase. As per the CDC (2014) report, about 1% of the world population has autism spectrum disorder; its prevalence has increased by 6–15% each year from 2002 to 2010. Cerebral palsy is the most common of all childhood disabilities, affecting about 764,000 children and adults in the USA.

Developmental infirmities have consistently been a fundamental but predominantly yet to be addressed public health problem for children in developing countries. They are described as disabilities associated primarily with the functioning of the neurological system and brain. There is an extensive and heterogeneous list of neurodevelopmental disorders, namely attention-deficit/hyperactivity disorder (ADHD), autism, learning disabilities, intellectual disability, conduct disorders, cerebral palsy and impairments in vision and hearing (Fig. 5.1). Children and adults affected by such developmental conditions are characterized by obstructed motor skills, speech and language impediment, behavioural abnormalities and difficulties with memory and learning or other neurological functions.

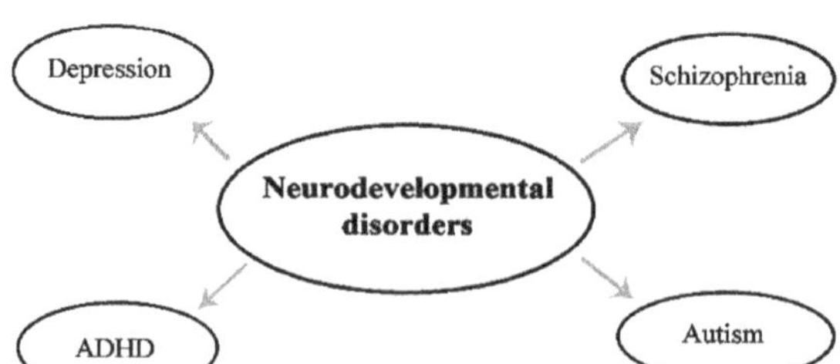

Figure 5.1 Neurodevelopmental disorders.

Neuropsychiatric syndromes are cerebral disorders that show coexistence of various psychiatric symptoms, such as:

- cognitive impairment;
- disturbance of consciousness;
- anxiety, neurotic complaint, apathy, mood disorder;
- hallucinations, delusions, behavioural differences;
- personality changes, delirium.

A symbiotic composition

The progress of growth in human beings has incorporated a symbiotic relationship of mutualism with the trillions of microbes residing in the intestinal tract contributing to a colossal phenotypic diversity and immensely shaping the host physiology and homeostatic mechanism. This includes close to over 1000 different species with more than 3 million genes (Qin *et al.*, 2010). The total colonic weight of these microorganisms is equivalent to 10^{11}–10^{12} cells/g, consisting of a biomass of 1–2 kg (Ley *et al.*, 2006; Stilling *et al.*, 2014). The colonization of microorganisms is distinctive for each individual determined by age, body mass index (BMI), medical history, lifestyle habits, geographical location, diet, antibiotic usage and host-related factors. Within the body there is a shift in the framework of microbes in each region, differing in the constitution of stomach region, skin, oral cavity, vagina and other intestinal regions. Predominantly anaerobes make up the population of microbes in the host, of which major proportions of phyla include Bacteroidetes, Firmicutes and Actinobacteria (Landman and Quévrain, 2016) and negligible proportions of Fusobacteria, Cyanobacteria, Proteobacteria and Verrucomicrobia are present. Throughout the lifespan of an individual, the composition of the intestinal microbes undergoes vigorous process of development and establishes a harmonious relationship with the host early in life. The infant gut flora acquisition begins during and after birth; apparently gestational age and mode of delivery (vaginal birth vs. assisted delivery) immensely influences the composition (Rutayisire *et al.*, 2016). The first microbial inoculum formed in the infant is secured from the maternal microbiota and later converges towards an individually distinct adult microbial profile. The ecosystem of the symbionts in neonates is characterized by low diversity and a relative supremacy of the phyla Enterobacteria, Staphylococci, Proteobacteria and Actinobacteria, taking advantage of the availability of oxygen in the newborn gut. Postnatal microbiotas become more diverse and earliest colonizers are outnumbered with the appearance and pre-eminence of *Bifidobacterium*, Firmicutes and Bacteroidetes. Various researchers have reported that these minute creatures in the gut perform a comprehensive array of functions referred to as microflora-associated characteristics (MAC). Among the several functions, the important ones include the host defence mechanism, production of bioactive compounds such as vitamin K and B, synthesis of neuroactive compounds, regulation of inflammatory response, digestive health, breakdown of complex carbohydrates and plant polyphenols, drug metabolism, mucosal cell development and stimulation of immune system. They have a direct relationship with the immune system and are essential in governing the principal functions of the immune system (Watzl, 2008). The gut indwelling commensal microorganisms have the inherent ability to provide protection against the invading pathogenic organisms via interaction with various antimicrobial proteins and peptides produced by the gut epithelial cells.

Cathelicidins, C-type lectins, (pro)defensins and histatins are some of the most important antimicrobial proteins, which strongly bind to the specific pattern recognition receptors (PRRs) and manifest their microbicidal, chemotactic activities and apoptotic effects targeting the infectious pathogenic microbes (Zhang and Gallo, 2016). The epithelial lining of the GI tract behaves as a protective barrier, whereby the specialized group of cells residing in the gut, namely the Paneth cells, provokes the innate immunity response by orchestrating the production of several antimicrobial proteins such as α-defensins (Sakai and Ayabe, 2008).

The colonization of the intestine runs in parallel with the immune system maturation and plays a crucial role in intestinal physiology and regulation. The interplay between the diverse microbial population and the host takes place through the immune system and secretion of numerous metabolites such as short-chain fatty acids (SCFAs) such as butyrate, propionate, acetate and valerate determines the physiological processes throughout the body. The bidirectional communication between the host and gut encompasses the supreme central nervous system (CNS) and enteric nervous system (ENS) as well as the neuroendocrine and neuroimmune system. In the life cycle of a neonate, microbial colonization of the intestine is a vital process as it potentially impacts the development of the brain. Adolescence is the most crucial period when the microbial composition is inclined to change. There are critical development windows indicating periods sensitive to change and characterized by dynamic changes in neuronal and microbial organization. The human microbiome is responsible for the secretion of various metabolites, hormones and neurotransmitters that impact the development and function of the brain (Autry and Monteggia, 2012).

The rapid growth of the microbes in the host GI tract relies on the diet, lifestyle, pH, neurotransmitters and gut physical–intestinal movements. The morphology of the gut is highly responsive to the dietary changes. The quality of food intake determines the richness of the microflora. The gut has high plasticity, which enables it to habituate to the changing environmental factors (Gussekloo, 2008). The thriving microflora require different kinds of nutrients for their proliferation, namely fibre-rich foods such as wholegrain s, polysaccharides, dietary polyphenols such as almonds, green tea, blueberries, sugars, lipids, proteins from the host intestinal tract, enterocyte tissues, bile acids and cholesterol. Change in the diet can bring about a major transformation in the gut microbiota. The disruptive factors that impact the host's normal flora include the use of antibiotics, immune suppression, stress, pathological condition and surgery.

The mind–gut intercommunication

A complex hierarchical intercommunication system exists between the gut and the brain axis, which prevails to execute high-level functions including nutrient metabolism, orchestrating brain development, physiological mechanisms and regulation of the interacting pathways. With the emergence of intestinal microbiota–brain axis, specific emphasis has paved the way for understanding the implication of microbiota in the maintenance of homeostasis. The potential unifying mechanism linking the brain–gut–immune system appears through various messenger molecules including neurotransmitters, neuropeptides, endocrine hormones and cytokines. The crosstalk between the intestinal flora and the brain includes an intricate interaction between the ENS, CNS and immune system (Mayer, 2011).

The vagus nerve, anatomically referred to as the tenth cranial nerve (X), accommodates the motor and sensory fibres. It represents the longest cranial nerve, which extends from the medulla oblongata vertically downwards to the GI tract. The right and left vagus nerve radiating from the brain stem branches outwards supplying various target organs via the pharyngeal nerve, superior laryngeal nerve, inferior cervical cardiac branch, thoracic cardiac branches, branches to the pulmonary plexus, and branches to the oesophageal plexus. Stimulation and activation of the afferent vagus nerve transmits information from the ENS to the brain circuits (Forsythe *et al.*, 2014). There is an abundance of neurotransmitters in the lumen of the GI tract, which are potent modulators of the activity of vagus nerve, namely acetylcholine and gamma-aminobutyric acid (GABA).

The bacterial metabolites produced in the gut as a result of fermentation have the capability to stimulate the sympathetic nervous system and influence production of neurotransmitters. These short-chain fatty acids including acetate, butyrate and propionate via the circulatory system can permeate the blood–brain barrier (BBB) and regulate the neuronal metabolic function and are involved in modulation of the other neurological functions. An altered expression of these metabolites has been implicated in the manifestation of various neurodevelopmental disorders. (Michel and Prat, 2016).

The immune system is the principal gatekeeper of the bidirectional signalling of the microbiota–brain axis. The constituents of the immune system such as cells, tissues and organs are intricately unified. The adaptive immune system and innate immune system elaborately mediate the two-way link between the intestinal microbes, alimentary tract and CNS (Lima-Ojeda *et al.*, 2017). The immune T cells, B cells, macrophages and dendritic cells send immune cytokine signals to the brain. The human brain is greatly fortified by the BBB preventing the entry of foreign pathogens. On the impulse of stimuli, various cytokines including interleukin (IL) 10, IL-6, IL-2, tumour necrosis factor-alpha (TNF-a), IL-1b can permeate the BBB resulting in activation of microglia and astrocytes near the brain capillaries. Microglia and astrocytes majorly contribute to the neuroimmune system, regulation of transport of neurotransmitters and ions, neurogenesis, synaptogenesis and homeostasis of the neuronal activity necessitates the requirement of these neuroglial cells (Bitzer-Quintero and González-Burgos, 2012). The hypothalamic–pituitary–adrenal (HPA) axis comprises the integration of the hypothalamus, endocrine glands and pituitary gland. It is the fundamental system for adaptation of the host physiological response to stress. The symbiotic bacterial organisms modulate the HPA axis via various signalling pathways. Various neurotransmitters released by the bacterial species modulate brain functioning, development and behaviour (Table 5.1). Deregulation of the level of neurotransmitters is attributed to the imbalance in gut microbiota. During fetal development, the gut microbiota determines the levels of serotonin (5-HT, 5-hydroxytryptamine) and brain-derived neurotrophic factor (BDNF) in regions of the amygdala and hippocampus (Sampson and Mazmanian, 2015). These factors are essentially involved in the contribution to neurogenesis and synaptic plasticity.

Brain-derived neurotrophic factor, commonly referred to as 'neurotrophin', belongs to the nerve growth factor family of proteins. It is a highly important protein influencing the sensory and sympathetic neurons. It is specifically required for the stimulation of synaptogenesis, neurogenesis, promotion of axonal growth of sensory neurons and support survival of existing neurons. Neurological conditions such as schizophrenia, depression, Alzheimer's disease and mood disorders such as anxiety and epilepsy show a decreased level of BDNF (Dinan and Cryan, 2013). Serotonin is a monoamine neurotransmitter acquired from the amino acid tryptophan. It is responsible for well-being, regulation of mood, sleep

Table 5.1 Specific neurotransmitters produced by different bacterial species

Bacterial species	Neurotransmitters
Lactobacillus spp.	GABA, acetylcholine
Bifidobacterium spp.	GABA
Escherichia spp.	Noradrenaline, serotonin
Bacillus spp.	Noradrenaline, dopamine
Candida spp.	Serotonin
Streptococcus spp.	Serotonin
Saccharomyces spp.	Noradrenaline

and cognitive functions. A deficit of serotonin causes depression. Serotonin synthesis requires tryptophan and this is generated by the intestinal microbiota, which can cross the BBB for the synthesis.

Enterochromaffin cells produce neurotransmitters, enteric neurons and glia under the influence of the commensal microorganisms. Particular microbes like *Clostridium sporogenes* and *Ruminococccus gnavus* residing in the gut provoke decarboxylation of the amino acid tryptophan by the neurotransmitter tryptamine (Williams *et al.*, 2014). Tryptamine includes a group of compounds such as melatonin and serotonin, which share the indole ring structure and side chains. This monoamine alkaloid is a non-selective serotonin receptor agonist that plays an inhibitory role in response to serotonin and regulates mood and appetite in the brain (Zucchi *et al.*, 2006). In the gut it has an opposite effect where it stimulates the enterochromaffin cells to secrete serotonin. Interestingly, studies have shown that numerous signals derived from the inhabiting microbes can also promote non-inflammatory cytokine pathways. For instance, the plasma granulocyte colony-stimulating factor (G-CSF) can traverse the BBB and acts to stimulate neurogenesis in the brain (Zhao *et al.*, 2007). Through G-CSF production the microbiota influences the rate of neurogenesis.

Neurochemicals and neuropeptides

The development process of the CNS requires a complex well-orchestrated series of events including neurogenesis, synaptogenesis, axonal and dendritic growth and interrelation of these pathways. At all stages of these events, gut microbes and their metabolites have been steadily reported to influence the growth and development. A well-functioning GI tract is imperative for a healthy nervous system. An imbalance in the gut flora provokes the synthesis of various toxic and inflammatory compounds, which inherently trigger the brain and are known to exacerbate neurological disorders. An instance of dysbiosis promotes various enzymes, neurotoxic metabolites such as D-lactic acid, ammonia and other metabolites to exhibit neurotoxicity. Cutting-edge research has recently explored the field of neurogastroenterology to gain a better understanding of the effects of neuromodulators released by the microbial community via various routes on the functioning of the nervous system.

Scientists have reported that about 90% of serotonin, the so-called 'feel-good' hormone, is produced in the GI tract. Enterochromaffin (EC) cells, neural cells and certain immune cells have the ability to release peripheral serotonin in the digestive tract. These EC cells are a copious source of serotonin and they rely on the diverse microbial populations for their production. Previous work undertaken in the field has measured the serotonin levels in mice containing normal populations of gut bacteria and also in germ-free mice. This concluded that the interaction of enterochromaffin cells with the varied species of gut microbes is necessary for the release of serotonin (Yano *et al.*, 2015). This neurotransmitter is responsible for the regulation of anxiety, stress, mood, neuron activity, social behaviour and maintenance of homeostasis of the brain function.

L-Glutamate and GABA possess excitatory and inhibitory roles as neurotransmitters, respectively. The biochemical relationship between glutamate and GABA maintains the regulatory networks involved in energy metabolism of neurons (Soeiro-de-Souza *et al.*, 2015). GABA is a principal inhibitory neurotransmitter present in the CNS, known to regulate pain and inflammation, and affects the personality and stress management. Various neural cells, immune cells and macrophages contain GABA receptors. The GABA channel receptors have the ability to change their shape on allowing specific negatively charged ions to pass through their central channel. About 40% of synapses interact with GABA and their receptors to exert their action on the brain.

Substance P, calcitonin gene-related peptide and neuropeptide Y (NPY) are important neuropeptides that have been mentioned to significantly contribute to the gut–brain signalling. Certain microbes in the gut are able to produce these neuropeptide-like molecules, owing to which any alterations in the microbial community may have an effect on the neuropeptide system in the CNS. The neuropeptides elicit signals to the brain by activating the primary afferent neurons, especially the vagus nerve besides the endocrine route. NPY is abundantly expressed at all levels of the gut–brain axis (Holzer *et al.*, 2012). Enteric neurons are the major source of expression of NPY within the gut; NPY plays a major role in several crucial pathways and functions including neurobiological homeostasis, immune function, synaptogenesis, expansion of interstitial neurons of the white matter and oligodendrocyte development. Within the brain they are highly constituted in the hypothalamus, hippocampus, cerebral cortex, basal ganglia, ventrolateral medulla, periaqueductal grey and locus coeruleus and amygdala (Wettstein *et al.*, 1995). Manifestation of aberrant GABA expression has been observed in many study cases of nervous system diseases with characteristic white matter abnormalities, such as anoxic–ischaemic injury, periventricular leukomalacia and schizophrenia (Wu and Sun, 2015).

Early exposure to gut microbes substantially contributes to the brain development and behaviour. The early years of life, which involve the maternal–fetal interaction, largely contribute to the dynamics of the gut microbiome. A signal released from the maternal microbiome is significant in fetal brain development (Funkhouser and Bordenstein, 2013). During synaptogenesis in the critical developmental period, it is conceivable that gut microbiota have the capacity to regulate synaptophysin and PSD-95 [postsynaptic density protein 95, also known as SAP-90 synapse-associated protein 90 (Diaz Heijtz *et al.*, 2011; Meyer *et al.*, 2014)]. Synaptophysin is a synaptic vesicle glycoprotein, an indirect marker of synaptogenesis in brain development. PSD-95 is a member of the membrane-associated guanylate kinase (MAGUK) family, which is an important protein product of the DLG4 (discs large homologue 4) gene. The principal function is to maintain synaptic plasticity and stabilization of synaptic changes during long-term potentiation. Aberrant PSD-95 expression has been highly mentioned in the emergence of Alzheimer's disease, ageing and several psychiatric disorders. Modulation of these synaptogenesis-associated proteins in the perinatal period has a long term effect on motor control and anxiety-like behaviour in adult life. The perinatal period is highly crucial for the developmental processes, as the gut microbes have the propensity to modulate them, namely modification of the HPA axis before six weeks of age post birth.

Modifying aspects of the functional gut

Several factors are known to influence the quality and composition of the microbial community inhabiting the gut; among them age, diet, urban lifestyle, mode of delivery, hospitalization, maternal disease, environment, genetics and antibiotic exposure are the major key players.

Age

From the time of nascency until one attains maturity the gut microbiota is constantly evolving within the lifespan of an individual. Mode of birth, type of feeding and maternal microbiome governs the initial newborn microbial composition. There is a significant displacement in the gut population after 2–3 years of birth, as influenced by the environmental exposure. The patterns and transition points in gut microbiota composition during childhood have been revealed to influence the health conditions later in life. Adolescence is the most critical period of development; any inhibition to the bacterial load of the gut can impact the development of the brain and may contribute to the manifestation of various psychiatric illnesses associated with puberty (Casey *et al.*, 2008; O'Connor and Cryan, 2014). Post adolescence the gut composition emerges towards adulthood and is found to be more stable. However, it begins to decline at old age; when the physiological functions of the human body deteriorate it enforces such dramatic changes. Immunosenescence, defined as a decline in the functionality of the immune system, is a major negative consequence of ageing, results in a chronic low-grade inflammatory status in the gut and therefore causes unfavourable changes in the composition and structure of the gut microbiota in older people (Gruver *et al.*, 2007).

Antibiotic usage

Antibiotic therapy is the best choice of treatment to extensively eliminate a broad category of bacterial infections such as respiratory tract infections, urinary tract infections, skin infections and infected wounds. However, a disturbing consequence of antibiotic treatment has been the long-term effects on the composition, ecology and resistance of the human gut microbiota. The systemic use of antibiotics alters the gut microbiota composition and disrupts the intestinal homeostasis. Panda *et al* (2014) reported the effects of ciprofloxacin and beta-lactams administered for a period of 7 days. They observed a reduction in the heterogeneity of the indigenous organisms by 25% and the core taxa was also decreased from 29 to 12, accompanied with an escalated Bacteroidetes-to-Firmicutes ratio. Several factors determine the magnitude of antibiotic-induced alterations in the gut, namely the spectrum of the agent, dosage and duration of treatment, route of administration and the pharmacokinetic and pharmacodynamic properties of the agent. Various high-throughput sequencing techniques such as quantitative polymerase chain reaction (qPCR) and 454 pyrosequencing have been implemented to analyse the effects of antibiotics on the structure of the microbial community. Recent studies have demonstrated the impact of antibiotics such as clarithromycin, metronidazole and omeprazole administered for the treatment of

Helicobacter pylori infections in dramatically perturbing the gut microbiota populations (Jakobsson *et al.*, 2010).

Diet

Diet is a major factor that shapes the composition and activity of the trillions of microorganisms residing in the gut throughout the life of an individual. A fibre-rich diet yields substrates for the intestinal microbial metabolism, as bacteria are particularly involved in the fermentation of different substrates resulting in SCFA production. Furthermore, increased consumption of wholegrain foods, fruits, vegetable, legumes and nuts positively influence the nutritional status of the gut. The proportion of the *Lactobacillus/Enterococcus* group was increased during a diet intervention period with wholegrain breakfast cereal. Similarly, a 3-week controlled study showed increased bifidobacteria levels in faeces of the volunteers on consumption of wholegrain breakfast cereal and placebo cereals (Carvalho-Wells *et al.*, 2010).

Maternal disease

The physiological and pathological features of pregnancy are known to influence the bacterial community in the intestine of the newborn. A viral infection in pregnant women has been reported to increase the likelihood of developing autism, schizophrenia and other neurodevelopmental disorders in their children. Scientists have replicated this interconnection in an experimental murine model by simulating a viral infection in pregnant mice by the process of inoculating the animals with a synthetic double-stranded RNA called poly(I:C). Thereafter, the progeny of mice receiving poly(I:C) was found to exhibit abnormal social and repetitive behaviours, such as ignoring new cage mates and persistently digging.

Role of the gut microbiome in neurodevelopment

Human brain development is an intricate process that commences from the third week of gestation and protracting after birth through late adolescence. Environmental exposure and stimuli have an indispensable role in refining the activity of neuronal circuits (Stiles and Jernigan, 2010); in particular, the microbiota is potentially able to influence the modelling of neuronal circuits and synaptic levels. Accumulating evidence, especially many animal studies, have suggested the role of microbiota in the standard gross morphology and ultrastructure of amygdala and hippocampus (Luczynski *et al.*, 2016). Amygdala is a prominent brain region that is principally involved in quite a few anxiety disorders. It is part of the limbic system, positioned deep within the temporal lobes of the brain. This central region is known to integrate the body's response to fear as well as other emotional responses and is critical in regulating social behaviour. Intestinal microorganisms have the capacity to modify the behaviour in response to anxiety and stress stimuli and regulate important CNS molecular changes at a transcriptional level. Several genes in the amygdala region have been altered in varied neurological disease states. Notably, the commensal microorganisms control the expression of several genes present in the brain. Anxiety disorders associated with the hyperactivity of amygdala or smaller anatomical size is noted in obsessive compulsive disorder (OCD), post-traumatic stress disorders and social anxiety disorder.

Experimental investigations comparing germ-free mice with conventionally colonized mice have highlighted the significance of microbiota in regulation of the brain structure.

Maladaptive modifications in amygdala and hippocampal volume have been reported in many anxiety disorders in humans and with early-life stress imposed in rodents (Salm *et al.*, 2004; Irle *et al.*, 2010) In addition, a number of studies have shown the impact of stress and anxiety on dendritic morphology of hippocampal and basolateral amygdala neurons (Vyas *et al.*, 2006; Allsop *et al.*, 2014). There is a complex relationship between the HPA axis signalling and amygdala, hippocampal activity and social behaviours. Recent evidence has shown that during pregnancy the maternal microbiota influences the immune system function of the offspring and any prenatal stress could manipulate the microbial composition in both rodents and humans (Golubeva *et al.*, 2015.).The ENS is an autonomous part of the nervous system primarily derived from the neuroectodermal progenitors originated from neural crest cells during embryogenesis. It is composed of nearly 500 million neurons and works in conjunction with the brain in response to various stimuli. Homeostasis of the GI tract is essential for the molecular interactions among microbiota, enteric neurons and immune cells. A direct neurochemical pathway exists between the integrated ENS system and CNS system, which provides for the propagation of the microbial promoted signal to the brain (Furness *et al.*, 2014.).The enteric neurons and glial cells have been known to regulate peristalsis, epithelial secretion and intestinal blood flow in addition to motor and secretory functions and incessantly monitor the dynamic niche of the gut wall and its lumen. The gut microbes interact with the ENS system and play an influential role in the occurrence of brain disease. ENS dysfunction has been implicated in the aetiopathogenesis of neurodegenerative disorders such as autism spectrum disorder, amyotrophic lateral sclerosis, transmissible spongiform encephalopathies and Parkinson's disease. Altered behaviour and brain function has been observed in gnotobiotic mice; the lack of microbes has caused them to be isolated and unable to socialize with other mice, accompanied by neurochemical changes in the brain (Neufeld *et al.*, 2011). Several experimental investigations using animal models of different neurodevelopment disorders such as ADHD, schizophrenia, autism, cognitive disability and depression established that gut microbiota composition greatly determines the brain development and function (Fig. 5.2). The synthesized neurochemicals and neuroactive compounds in the intestinal lumen can traverse the BBB and induce many cognitive and behavioural disturbances (Wakefield, 2002). High-throughput analytical tools facilitate

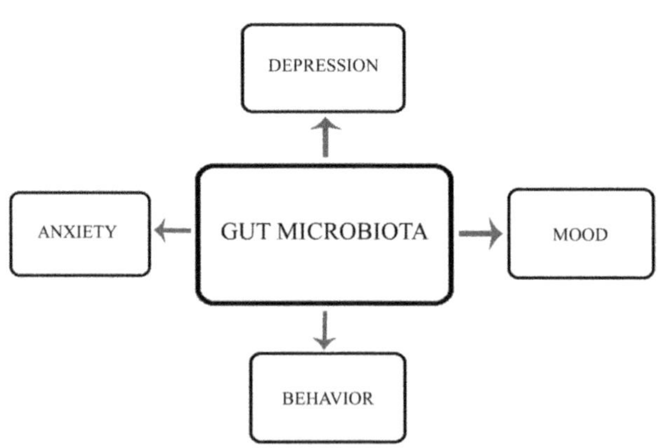

Figure 5.2 Effects of gut microbiota on the brain.

probing of the host–microbiota relationship at high resolution, helping to relate healthy or diseased states to the intricate composition of the microbiota.

Attention deficit hyperactivity disorder

The American Psychological Association defines ADHD as a neurobehavioural condition characterized by a persistent pattern of inattention and hyperactivity that interferes with the brain functioning and development. Over the years, it has become highly prevalent considering the lifestyle and diet change among urban populations, which principally transforms the microbial colonization of the intestine (Rogers *et al.*, 2016). It has been reported that about 5–12% of children show symptoms of ADHD worldwide and evidence from preliminary human studies has supported the link between ADHD traits and imbalanced gut microbiota composition (Polanczyk.*et al.*, 2007). The manifestation of the disorder is not limited to children alone but also continues in adulthood. There is a poor connectivity between different regions of the brain and dysfunction of the communication network interferes with the performance. A marked disruption in the communication of neurotransmitters linking the neurons impairs the stimulation of various brain functions. Alterations in the level of serotonin and the pathway involving dopamine activity have been mentioned to agitate the brain of ADHD individuals (Oades, 2008). Knowing that aberrant gut health provokes the occurrence of many neurological problems, thereby therapeutically targeting gut dysbiosis to manipulate the intestinal composition of microbial organisms in a way so that the 'good' bacteria are able to flourish and pathogens are suppressed, will be the focus of future treatment strategies for the majority of mental health problems.

Schizophrenia

Schizophrenia is an agonizing disorder of the mind and the brain, globally afflicting more than 21 million people. It is categorized among the top ten causes of mental disability in developed countries worldwide, irrespective of class, colour, race and religion. This chronic disorder is characterized by impaired functioning of the brain, which leads to disrupted thinking ability and loss of personality and perception. The most consistently supported hypothesis proposes that neurological and psychiatric disorders are triggered when environmental disturbances interact with genetic predispositions (McGrath *et al.*, 2008). This most likely occurs during fetal development or preliminary postpartum brain development when neurons are migrating and synapses are being initiated The integrity of the GI tract and GI motility functions have been recently studied for their relevance in a number of psychiatric disorders. The homeostatic microbial community is disrupted in adults suffering from schizophrenia and it tends to hasten the disease process. The aetiology and pathophysiology of schizophrenia involve the signalling effects of commensal gut microbes on the mechanism and behaviour of the CNS. Several studies have probed the apparent association found between the gut microbiome and behavioural response in schizophrenic patients; notably, analysis in mice has shown that microbiota have the capacity to modulate the programming of social behaviours and cognition traits known to be impaired in schizophrenia (Desbonnet *et al.*, 2014). Preclinical animal model studies have demonstrated the regulatory role of gut microbiota in dysfunctional brain, anxiety disorders and depression-related behaviour; the vagus nerve seems to be a potent arbitrator in these associations.

Clinical human studies have suggested a role for alterations in the gut and gut microbiota in schizophrenia, as perturbation of inflammatory processes has been implicated in schizophrenia (Al-Asmari *et al.*, 2014). The gut microbiotas are hypothesized to regulate immune processes and maintain the epithelial barrier integrity (Lopetuso *et al.*, 2014), owing to which it has been suggested that the microbiome–gut–brain axis could influence schizophrenia through immunological mechanisms. Inflammation, infection and stress are environmental factors that compromise the microbial populations consequently, induce toxic and bioactive products derived from the gut to escape the GI tract and produce an immune response and then is transmitted to the brain. A heightened threat of schizophrenia (Benros *et al.*, 2011) and mood disorders (Benros *et al.*, 2013) has been indicated in various autoimmune diseases characterized by chronic gastrointestinal inflammation, such as coeliac disease, Crohn's disease and ulcerative colitis. Schwarz *et al.* (2017) examined indicative variations in the composition of faecal microbiota between 28 patients with first-episode psychosis (FEP) and 16 healthy matched controls and analysed whether such differences were associated with the response after up to 12 months of treatment. Patients with FEP and controls were well matched with respect to age, body mass index (BMI), sex and markers of glucose and lipid metabolism. The study concluded that about 50% of patients received a diagnosis of schizophrenia at the 1-year follow-up and displayed poor functioning (Schwarz *et al.*, 2017). Further, the population of the commensal group of bacteria, namely *Lactobacillus*, was correlated positively with the gravity of psychotic symptoms and negatively with global assessment of functioning. In conclusion, the available evidence supports the participation of gut dysbiosis in the development of psychiatric illness and may lay the foundation for investigating the advantageous gain of their modulation on treatment response and remission.

Depression

Depression is a serious mental disorder that occurs worldwide. It is characterized by persistent sadness, anxiety and irritability, reduced energy leading to diminished activity and disturbed sleep and eating habits. This common mental illness is widespread, affecting more than 300 million people, irrespective of age, status and nationality (Johnson *et al.*, 1992). A distinct imbalance in the structure of the intestinal ecosystem of microbes was manifested in murine models of depression and chronic stress. This has led to the development of psychobiotic-based therapeutic strategies for various neurological disorders. Depression is the foremost cause of disability worldwide and contributes to the overall morbidity and mortality rate. Severe and recurrent depression can lead to the occurrence of bipolar disorder, which typically includes both manic and depressive episodes separated by periods of a well-balanced normal mood characterized by mood fluctuations, hyperactivity, pressure of speech, inflated self-esteem and insomnia.

Recent neuroscientific research has demonstrated the significance of the microbiota in the development of brain systems and activation of the immune system and CNS. Oral administration of *Bifidobacterium infantis to* germ-free mice was found to reverse the hyperactive HPA axis response and depression state as well as resulting in relaxation of neuronal processes (Desbonnet *et al.*, 2010). Gut microorganisms have the propensity of synthesizing and secreting neuroactive substances such as serotonin and GABA, which act on the gut–brain axis. Maladaptive stimulation of the immune system could result in the release of several pro-inflammatory cytokines and chemokines, such as interleukin (IL)-8, and IL-1, IL-10 and transforming growth factor-beta (TGF-β) and interferon-alpha are known to induce depression. It is quite well known that persistent exposure to increased levels of inflammatory cytokines and frequent alterations in neurotransmitter response systems can lead to neuropsychiatric disorders and depression. It disrupts the physiological functioning of the entire body and fails to maintain homeostasis. Mechanisms underlying cytokine behavioural effects involve accelerated inflammatory signalling pathways in the brain that provoke variations in monoamine, glutamate, and neuropeptide systems and rapidly reduce growth factors, e.g. BDNF.

Definite alteration in the stress response and overall behaviour is associated with the corresponding modifications in the brain–gut axis function. Stress triggers a chemical change in the brain that modifies the synapses, thereby disturbing the neuronal connectivity. During the period of exposure to stress, there are elevated levels of cortisol, commonly referred to as 'stress hormone', which is the body's natural response to stress. Response to stress is an essential part of survival. However, while enabling the body to manage the effects of stress, this hormone has devastating effects on the brain. The high levels of glucocorticoid activity are associated with reduced neurogenic activity (Gould *et al.*, 1997; Schoenfeld *et al.*, 2013). O'Mahony *et al.* (2011) evaluated the unfavourable outcome of neonatal stress on the functioning of the brain–gut axis. A group of male rats subsequent to birth were subjected to stress, owing to maternal separation for about three hours daily at between 2 and 12 days, while the control group was left undisturbed with their mothers. Potential parameters indicating significant alterations including behaviour, immune response, stress sensitivity, visceral sensation and faecal microbiota were inspected. The number of faecal boli deposits increased in response to early life stress (O'Mahony *et al.*, 2011). A consequential rise in the level of the principal stress hormone corticosterone was observed in the maternally separated animals by the time they reached adulthood.

Prolonged stress is known to alter the BBB, permitting the entry of pathogens, heavy metal and toxins, which damage the brain cells. There is an immense loss of neurons and brain cells due to the inhibitory action of cortisol on BDNF, which is a protein used to maintain healthy brain cells (Yulug *et al.*,2008). Therefore, modulation of the brain–gut axis is being seen as an attractive target for the development of novel treatments for a wide variety of disorders ranging from autism to schizophrenia, depression, Alzheimer's disease and anxiety.

Prospective therapeutic interventions

The scope of probiotics and prebiotics in conferring good health to the host has enlightened the minds of many ardent scientists and researchers. Aware of the compelling participation of gut microbes in the multifold complex host functions, care should be taken to nutrify the innate gut microbial populations. They aid in restoring the normal microflora post the incidence of severe dysbiosis of the gut (Levin, 2011). Additionally, they help to ameliorate the manifestations of the imbalanced gut–brain communication. The probiotics are designed and formulated specifically for particular diseases and they are shown to exhibit antimicrobial, anticancer and antitumour effects (Singh *et al.*, 2017). Although the undoing effects of stress on the physiology of the brain and general body function are known, there is an increased permeability of the gut, which mediates various pathologies (Pathak *et al.*, 2017). The clinical effects of a symbiotic preparation of *Saccharomyces boulardi* have been investigated to target the negative effects of mild stress instigated intestinal hyperpermeability (Takadanohara *et al.*, 2012). Exploiting high-throughput techniques such as 16S rRNA sequencing and shotgun metagenomics help in attaining complete understanding of the composition, structure and functioning of the gut microbiota, which is pivotal in addressing disease-related treatment modalities (Nagpal *et al.*, 2014).

Future outlook and conclusion

The human brain is a remarkably complex organ whose development commences in the third gestational week with the differentiation of the neural progenitor cells and sustains at least through late adolescence, perhaps throughout the lifespan. A growing body of experimental data and clinical observations has unveiled the unexpected link between the microbiota–gut–brain axis and suggests that it is poised to control the canonical aspects of brain development and overall mental health. In particular, the prenatal, postnatal and adolescence phases are the critical periods of interacting microbial–neural systems. The gut microbiota plays an influential role not just in normal brain development; various aspects such as cognitive function, behaviour, mood, social interaction are dependent on the diversity and structure of the gut microbiota. In the past decade much effort has been dedicated in understanding how neurophysiology can be manipulated by the symbiotic microbial community and how alterations in the gut commensals during critical periods of postnatal growth and differentiation of new neural structures and functions might impact neurodevelopmental trajectories. Several studies have demonstrated the consequential effects of a perturbed gut microbiome; this has contributed to the occurrence of various neurodevelopmental and psychiatric disorders such as autism, schizophrenia, Alzheimer's disease, Parkinson's disease, cognitive impairment, stress-related conditions, anxiety and depression. Unravelling the various determinants that are responsible for disrupting the gut homeostasis will improve our knowledge of the aetiology of neurodevelopmental psychiatric disorders and enable us to identify new therapeutic approaches to neuropathological symptoms of various behavioural disorders and their comorbid medical illnesses. Ongoing research may potentially improve our understanding of the significance of early interactions between the intestinal microbiota and the host CNS as well as identifying the microbial metabolites that can be used to better monitor brain inflammation and the behavioural response. In conclusion, this chapter advocates the need for further research in this area with the perception that a balance in the gut microbiota would facilitate the improvement of the treatment of neuropsychiatric diseases from bench to the bedside.

Acknowledgements

The authors are thankful to the Chettinad Academy of Research and Education (CARE), Chennai, India, for the research facilities and support.

Funding

This study was partially financially supported by grants from the Chettinad Hospital and Research Institute (CHRI), Chettinad Academy of Research and Education (CARE), Chennai, India.

Conflicts of interests

No potential conflict of interest was reported by the authors.

Author contribution

SVP, SP, AB wrote the manuscript. SP, RM and FM designed and revised the manuscript.

References

Allsop, S.A., Vander Weele, C.M., Wichmann, R., and Tye, K.M. (2014). Optogenetic insights on the relationship between anxiety-related behaviors and social deficits. Front. Behav. Neurosci. *8*, 241. https://doi.org/10.3389/fnbeh.2014.00241.

Al-Asmari, A.K., and Khan, M.W. (2014). Inflammation and schizophrenia: Alterations in cytokine levels and perturbation in antioxidative defense systems. Hum. Exp. Toxicol. *33*, 115–122. https://doi.org/10.1177/0960327113493305.

Autry, A.E., and Monteggia, L.M. (2012). Brain-derived neurotrophic factor and neuropsychiatric disorders. Pharmacol. Rev. *64*, 238–258. https://doi.org/10.1124/pr.111.005108.

Benros, M.E., Nielsen, P.R., Nordentoft, M., Eaton, W.W., Dalton, S.O., and Mortensen, P.B. (2011). Autoimmune diseases and severe infections as risk factors for schizophrenia: A 30-year population-based register study. Am. J. Psychiatry *168*, 1303–1310. https://doi.org/10.1176/appi.ajp.2011.11030516.

Benros, M.E., Waltoft, B.L., Nordentoft, M., Ostergaard, S.D., Eaton, W.W., Krogh, J., and Mortensen, P.B. (2013). Autoimmune diseases and severe infections as risk factors for mood disorders: A nationwide study. JAMA Psychiatry *70*, 812–820. https://doi.org/10.1001/jamapsychiatry.2013.1111.

Bitzer-Quintero, O.K., and González-Burgos, I. (2012). Immune system in the brain: A modulatory role on dendritic spine morphophysiology? Neural Plast. *2012*, 348642. https://doi.org/10.1155/2012/348642.

Carvalho-Wells, A.L., Helmolz, K., Nodet, C., Molzer, C., Leonard, C., McKevith, B., Thielecke, F., Jackson, K.G., and Tuohy, K.M. (2010). Determination of the *in vivo* prebiotic potential of a maize-based whole grain breakfast cereal: a human feeding study. Br. J. Nutr. *104*, 1353–1356. https://doi.org/10.1017/S0007114510002084.

Casey, B.J., Jones, R.M., and Hare, T.A. (2008). The adolescent brain. Ann. N. Y. Acad. Sci. *1124*, 111–126. https://doi.org/10.1016/j.dr.2007.08.003.

Desbonnet, L., Clarke, G., Shanahan, F., Dinan, T.G., and Cryan, J.F. (2014). Microbiota is essential for social development in the mouse. Mol. Psychiatry *19*, 146–148. https://doi.org/10.1038/mp.2013.65.

Desbonnet, L., Garrett, L., Clarke, G., Kiely, B., Cryan, J.F., and Dinan, T.G. (2010). Effects of the probiotic *Bifidobacterium infantis* in the maternal separation model of depression. Neuroscience *170*, 1179–1188. https://doi.org/10.1016/j.neuroscience.2010.08.005.

Diaz Heijtz, R., Wang, S., Anuar, F., Qian, Y., Björkholm, B., Samuelsson, A., Hibberd, M.L., Forssberg, H., and Pettersson, S. (2011). Normal gut microbiota modulates brain development and behavior. Proc. Natl. Acad. Sci. U.S.A. *108*, 3047–3052. https://doi.org/10.1073/pnas.1010529108.

Dinan, T.G., and Cryan, J.F. (2013). Melancholic microbes: A link between gut microbiota and depression? Neurogastroenterol. Motil. *25*, 713–719. https://doi.org/10.1111/nmo.12198.

Forsythe, P., Bienenstock, J., and Kunze, W.A. (2014). Vagal pathways for microbiome-brain-gut axis communication. In Microbial Endocrinology: The Microbiota-Gut-Brain Axis in Health and Disease, Lyte, M., and Cryan, J.F., eds., Adv. Exp. Med. Biol. *817*, 115–133. https://dx.doi.org/10.1007/978-1-4939-0897-4_5.

Funkhouser, L.J., and Bordenstein, S.R. (2013). Mom knows best: The universality of maternal microbial transmission. PLOS Biol. *11*, e1001631. https://doi.org/10.1371/journal.pbio.1001631.

Furness, J.B., Callaghan, B.P., Rivera, L.R., and Cho, H.J. (2014). The enteric nervous system and gastrointestinal innervation: Integrated local and central control. Adv. Exp. Med. Biol. *817*, 39–71. https://doi.org/10.1007/978-1-4939-0897-4_3.

Golubeva, A.V., Crampton, S., Desbonnet, L., Edge, D., O'Sullivan, O., Lomasney, K.W., Zhdanov, A.V., Crispie, F., Moloney, R.D., Borre, Y.E., *et al.* (2015). Prenatal stress-induced alterations in major physiological systems correlate with gut microbiota composition in adulthood. Psychoneuroendocrinology *60*, 58–74. https://doi.org/10.1016/j.psyneuen.2015.06.002.

Gould, E., McEwen, B.S., Tanapat, P., Galea, L.A., and Fuchs, E. (1997). Neurogenesis in the dentate gyrus of the adult tree shrew is regulated by psychosocial stress and NMDA receptor activation. J. Neurosci. *17*, 2492–2498.

Gruver, A.L., Hudson, L.L., and Sempowski, G.D. (2007). Immunosenescence of ageing. J. Pathol. *211*, 144–156. https://doi.org/10.1002/path.2104.

Gussekloo, S.W.S. (2008). Evolution of the gut: Adaptation and plasticity. Int. J. Probiotics Prebiotics *3*, 111–118.

Holzer, P., Reichmann, F., and Farzi, A. (2012). Neuropeptide Y, peptide YY and pancreatic polypeptide in the gut-brain axis. Neuropeptides *46*, 261–274. https://doi.org/10.1016/j.npep.2012.08.005.

Irle, E., Ruhleder, M., Lange, C., Seidler-Brandler, U., Salzer, S., Dechent, P., Weniger, G., Leibing, E., and Leichsenring, F. (2010). Reduced amygdalar and hippocampal size in adults with generalized social phobia. J. Psychiatry Neurosci. *35*, 126–131.

Jakobsson, H.E., Jernberg, C., Andersson, A.F., Sjölund-Karlsson, M., Jansson, J.K., and Engstrand, L. (2010). Short-term antibiotic treatment has differing long-term impacts on the human throat and gut microbiome. PLOS ONE *5*, e9836. https://doi.org/10.1371/journal.pone.0009836.

Johnson, J., Weissman, M.M., and Klerman, G.L. (1992). Service utilization and social morbidity associated with depressive symptoms in the community. JAMA *267*, 1478–1483.

Lagace, D.C., Donovan, M.H., DeCarolis, N.A., Farnbauch, L.A., Malhotra, S., Berton, O., Nestler, E.J., Krishnan, V., and Eisch, A.J. (2010). Adult hippocampal neurogenesis is functionally important for stress-induced social avoidance. Proc. Natl. Acad. Sci. U.S.A. *107*, 4436–4441. https://doi.org/10.1073/pnas.0910072107.

Landman, C., and Quévrain, E. (2016). [Gut microbiota: Description, role and pathophysiologic implications.] Rev. Med. Interne *37*, 418–423. https://doi.org/10.1016/j.revmed.2015.12.012.

Ley, R.E., Peterson, D.A., and Gordon, J.I. (2006). Ecological and evolutionary forces shaping microbial diversity in the human intestine. Cell *124*, 837–848.

Lima-Ojeda, J.M., Rupprecht, R., and Baghai, T.C. (2017). 'I am I and my bacterial circumstances': Linking gut microbiome, neurodevelopment, and depression. Front. Psychiatry *8*, 153. https://doi.org/10.3389/fpsyt.2017.00153.

Lopetuso, L.R., Felice, C., Pugliese, D., Scaldaferri, F., Gasbarrini, A., and Armuzzi, A. (2014). Innate immune response and gut microbiota in inflammatory bowel disease. CellR4 *2*, e1212.

Luczynski, P., McVey Neufeld, K.A., Oriach, C.S., Clarke, G., Dinan, T.G., and Cryan, J.F. (2016). Growing up in a bubble: Using germ-free animals to assess the influence of the gut microbiota on brain and behavior. Int. J. Neuropsychopharmacol. *19*, pyw020. https://doi.org/10.1093/ijnp/pyw020.

Mayer, E.A. (2011). Gut feelings: The emerging biology of gut-brain communication. Nat. Rev. Neurosci. *12*, 453–466. https://doi.org/10.1038/nrn3071.

McGrath, J., Saha, S., Chant, D., and Welham, J. (2008). Schizophrenia: A concise overview of incidence, prevalence, and mortality. Epidemiol. Rev. *30*, 67–76. https://doi.org/10.1093/epirev/mxn001.

Meyer, D., Bonhoeffer, T., and Scheuss, V. (2014). Balance and stability of synaptic structures during synaptic plasticity. Neuron *82*, 430–443. https://doi.org/10.1016/j.neuron.2014.02.031.

Michel, L., and Prat, A. (2016). One more role for the gut: Microbiota and blood brain barrier. Ann. Transl. Med. *4*, 15. https://doi.org/10.3978/j.issn.2305-5839.2015.10.16.

Neufeld, K.M., Kang, N., Bienenstock, J., and Foster, J.A. (2011). Reduced anxiety-like behavior and central neurochemical change in germ-free mice. Neurogastroenterol. Motil. *23*, 255–264, e119. https://doi.org/10.1111/j.1365-2982.2010.01620.x.

Nagpal, R., Yadav, H., and Marotta, F. (2014). Gut microbiota: The next-gen frontier in preventive and therapeutic medicine? Front. Med. *1*, 15. https://doi.org/10.3389/fmed.2014.00015.

Sakai, N., and Ayabe, T. (2008). Antimicrobial peptides in the gut innate immunity. Int. J. Probiotics Prebiotics *3*, 123–126.

Oades, R.D. (2008). Dopamine-serotonin interactions in attention-deficit hyperactivity disorder (ADHD). Prog. Brain Res. *172*, 543–565. https://doi.org/10.1016/S0079-6123(08)00926-6.

O'Connor, R.M., and Cryan, J.F. (2014). Adolescent brain vulnerability and psychopathology through the generations: Role of diet and dopamine. Biol. Psychiatry *75*, 4–6. https://doi.org/10.1016/j.biopsych.2013.10.022.

O'Mahony, S.M., Hyland, N.P., Dinan, T.G., and Cryan, J.F. (2011). Maternal separation as a model of brain-gut axis dysfunction. Psychopharmacology *214*, 71–88. https://doi.org/10.1007/s00213-010-2010-9.

Panda, S., El Khader, I., Casellas, F., López Vivancos, J., García Cors, M., Santiago, A., Cuenca, S., Guarner, F., and Manichanh, C. (2014). Short-term effect of antibiotics on human gut microbiota. PLOS ONE *9*, e95476. https://doi.org/10.1371/journal.pone.0095476.

Pathak, S., Suhanya, P., Sushmitha, S., Murugesan, R., He, F., Marotta, F., and Banerjee, A. (2017). Coping with stress related effects on the brain, role of neuro-nutraceuticals and gut microbes. Int. J. Probiotics Prebiotics *12*, 183–190.

Polanczyk, G., de Lima, M.S., Horta, B.L., Biederman, J., and Rohde, L.A. (2007). The worldwide prevalence of ADHD: A systematic review and metaregression analysis. Am. J. Psychiatry *164*, 942–948.

Qin, J., Li, R., Raes, J., Arumugam, M., Burgdorf, K.S., Manichanh, C., Nielsen, T., Pons, N., Levenez, F., Yamada, T., *et al.* (2010). A human gut microbial gene catalogue established by metagenomic sequencing. Nature *464*, 59–65. https://doi.org/10.1038/nature08821.

Rogers, G.B., Keating, D.J., Young, R.L., Wong, M.L., Licinio, J., and Wesselingh, S. (2016). From gut dysbiosis to altered brain function and mental illness: Mechanisms and pathways. Mol. Psychiatry *21*, 738–748. https://doi.org/10.1038/mp.2016.50.

Levin, R. (2011). Probiotics – the road map. Int. J. Probiotics Prebiotics *6*, 133–140.

Rutayisire, E., Huang, K., Liu, Y., and Tao, F. (2016). The mode of delivery affects the diversity and colonization pattern of the gut microbiota during the first year of infants' life: A systematic review. BMC Gastroenterol. *16*, 86. https://doi.org/10.1186/s12876-016-0498-0.

Salm, A.K., Pavelko, M., Krouse, E.M., Webster, W., Kraszpulski, M., and Birkle, D.L. (2004). Lateral amygdaloid nucleus expansion in adult rats is associated with exposure to prenatal stress. Brain Res. Dev. Brain Res. *148*, 159–167. https://doi.org/10.1016/j.devbrainres.2003.11.005.

Sampson, T.R., and Mazmanian, S.K. (2015). Control of brain development, function, and behavior by the microbiome. Cell Host Microbe *17*, 565–576. https://doi.org/10.1016/j.chom.2015.04.011.

Schwarz, E., Maukonen, J., Hyytiäinen, T., Kieseppä, T., Orešič, M., Sabunciyan, S., Mantere, O., Saarela, M., Yolken, R., and Suvisaari, J. (2017). Analysis of microbiota in first episode psychosis identifies preliminary associations with symptom severity and treatment response. Schizophr. Res. *192*, 398–403. https://doi.org/10.1016/j.schres.2017.04.017.

Schoenfeld, T.J., and Gould, E. (2013). Differential effects of stress and glucocorticoids on adult neurogenesis. Curr. Top. Behav. Neurosci. *15*, 139–164. https://doi.org/10.1007/7854_2012_233.

Singh, B., Mal, G., and Marotta, F. (2017). Designer probiotics: Paving the way to living therapeutics. Trends Biotechnol. *35*, 679–682.

Soeiro-de-Souza, M.G., Henning, A., Machado-Vieira, R., Moreno, R.A., Pastorello, B.F., da Costa Leite, C., Vallada, H., and Otaduy, M.C. (2015). Anterior cingulate Glutamate-Glutamine cycle metabolites are altered in euthymic bipolar I disorder. Eur. Neuropsychopharmacol. *25*, 2221–2229. https://doi.org/10.1016/j.euroneuro.2015.09.020.

Stiles, J., and Jernigan, T.L. (2010). The basics of brain development. Neuropsychol. Rev. *20*, 327–348. https://doi.org/10.1007/s11065-010-9148-4.

Stilling, R.M., Dinan, T.G., and Cryan, J.F. (2014). Microbial genes, brain & behaviour – epigenetic regulation of the gut-brain axis. Genes Brain Behav. *13*, 69–86. https://doi.org/10.1111/gbb.12109.

Takadanohara, H., Catanzaro, R., Chui, d.e.H., He, F., Yadav, H., Ganguli, A., Sakata, Y., Solimene, U., Minelli, E., Kobayashi, R., *et al.* (2012). Beneficial effect of a symbiotic preparation with *S. boulardii* lysate in mild stress-induced gut hyper-permeability. Acta Biomed. *83*, 208–216.

Vyas, A., Jadhav, S., and Chattarji, S. (2006). Prolonged behavioral stress enhances synaptic connectivity in the basolateral amygdala. Neuroscience *143*, 387–393.

Wakefield, A.J. (2002). The gut-brain axis in childhood developmental disorders. J. Pediatr. Gastroenterol. Nutr. *34* (Suppl. 1), S14–7.

Watzl, B. (2008). Bacterial metabolites and immune modulation. Int. J. Probiotics Prebiotics *3*, 133–136.

Wettstein, J.G., Earley, B., and Junien, J.L. (1995). Central nervous system pharmacology of neuropeptide Y. Pharmacol. Ther. *65*, 397–414.

Williams, B.B., Van Benschoten, A.H., Cimermancic, P., Donia, M.S., Zimmermann, M., Taketani, M., Ishihara, A., Kashyap, P.C., Fraser, J.S., and Fischbach, M.A. (2014). Discovery and characterization of gut microbiota decarboxylases that can produce the neurotransmitter tryptamine. Cell Host Microbe *16*, 495–503. https://doi.org/10.1016/j.chom.2014.09.001.

Wu, C., and Sun, D. (2015). GABA receptors in brain development, function, and injury. Metab. Brain Dis. *30*, 367–379. https://doi.org/10.1007/s11011-014-9560-1.

Yano, J.M., Yu, K., Donaldson, G.P., Shastri, G.G., Ann, P., Ma, L., Nagler, C.R., Ismagilov, R.F., Mazmanian, S.K., and Hsiao, E.Y. (2015). Indigenous bacteria from the gut microbiota regulate host serotonin biosynthesis. Cell *161*, 264–276. https://doi.org/10.1016/j.cell.2015.02.047.

Yuluğ, B., Ozan, E., Gönül, A.S., and Kilic, E. (2009). Brain-derived neurotrophic factor, stress and depression: A minireview. Brain Res. Bull. *78*, 267–269. https://doi.org/10.1016/j.brainresbull.2008.12.002.

Zhao, L.R., Navalitloha, Y., Singhal, S., Mehta, J., Piao, C.S., Guo, W.P., Kessler, J.A., and Groothuis, D.R. (2007). Hematopoietic growth factors pass through the blood-brain barrier in intact rats. Exp. Neurol. *204*, 569–573.

Zhang, L.J., and Gallo, R.L. (2016). Antimicrobial peptides. Curr. Biol. *26*, 14–19. https://doi.org/10.1016/j.cub.2015.11.017.

Zucchi, R., Chiellini, G., Scanlan, T.S., and Grandy, D.K. (2006). Trace amine-associated receptors and their ligands. Br. J. Pharmacol. *149*, 967–978.

An Insight into the Link between Oral Health and Neurological Diseases

6

Surajit Pathak[1]*, Suhanya Veronica Prasad[1], Sushmitha Sriramulu[1], Ganesan Jothimani[1], Ramachandran Murugesan[1], Francesco Marotta[2]* and Antara Banerjee[1]

[1]Faculty of Allied Health Sciences, Chettinad Academy of Research and Education, Kelambakkam, Chennai-603103, India.
[2]ReGenera R&D International for Aging Intervention, Milano-Beijing, and VCC Preventive Medical Promotion Foundation, Beijing, China.

*Correspondence: Surajit.pathak@gmail.com and fmarchimede@libero.it

https://doi.org/10.21775/9781910190937.06

Abstract

The oral cavity is host to a diverse population of microorganisms and it significantly contributes to the health of an individual. An imbalance or disruption to the oral ecosystem can lead to detrimental health complications. There are approximately 700 different species found, of which both mutualistic and pathogenic microbes are present. The most abundant genera in the mouth include *Streptococcus, Actinomyces, Fusobacterium, Haemophilus and Porphyromonas*. The human oral microbiome database (HOMD) was created to provide comprehensive information on the prokaryotic species. Currently, about 400 oral taxa have been sequenced. The oral mucosa, saliva and gingival crevicular fluid provide proteins that support microbial growth. The relationship between the oral microbiome and neurological diseases has attracted much attention over the recent years. Cases have been reported on the association of certain types of stroke and the presence of oral bacteria. Predominantly *Streptococcus* mutants are the causal agent of haemorrhagic stroke; these mutants are found in one-third of patients who have had stroke. This strain of bacteria is a major contributor of gum disease and tooth decay. It degrades the oral structure and attaches itself to vulnerable red blood cells and circulates to the brain, where it weakens the arterial walls reducing the blood flow and thereby causing stroke. Interestingly, scientific findings have shown the occurrence of oral pathogens in Alzheimer's disease patients, with about a 7-fold higher density than cognitively normal controls. The bacteria produce proteins and microbial toxins encoded by the genes that disrupt the blood–brain barrier inducing various complications. Local inflammatory responses involving the production of pro-inflammatory cytokines such as interleukin (IL) 6 and IL-1β have been implicated by these oral bacteria in the pathology of neurodegenerative diseases such as Parkinson's disease and Alzheimer's disease. Recently emerged next-generation sequencing and bioinformatics tools help to expand knowledge of the composition and function of the oral microbiome. Assessment of oral pathogens can be utilized as biomarkers for diagnosis of diseases. In this chapter, we will emphasize how oral microbiota cross the blood–brain barrier and can affect brain health.

Introduction

The human body shares an intricate relationship of codependency with the minute creatures residing in various habitats of the host, inclusive of bacteria, fungi, archaea and viruses. Understanding of the intestinal microflora has been appreciably explored and has received much recognition in the past decade, emphasizing their significance in health and disease. Besides the gut, various other locations including the skin, oral cavity, nasopharynx and urogenital tract are abundantly colonized. The mouth is the primary port of entry into the body, comprising the teeth, cheeks, lips, palate, tongue and gingival crevice, including subgingival and supragingival areas (Dewhirst *et al.*, 2010). The surfaces of these components are characterized by clusters of biofilm communities. Second to the gut, the oral cavity encompasses a varied population of microorganisms, of which 54% have been officially identified and cultivated. The mouth and the mind are not only interconnected anatomically; surreptitiously the oral microflora affects the central nervous system. Oral health care is essential for maintaining dental hygiene and preventing the occurrence of mouth lesions, dental carries, gum diseases, endodontic infection, alveolar osteitis and halitosis (bad breath). Beyond the evident health effects, the link between various health problems including neurological diseases, systemic diseases, nutritional malfunctions and poor oral hygiene has been recently unveiled. Recently updated genomic technologies, especially the 16S rRNA sequencing technique, have acquired the leverage to gain a comprehensive knowledge of the existing 700 prokaryote species in the oral cavity (Aas *et al.*, 2005). The colonization of microbes on various surfaces necessitates the formation of biofilms on the oral mucous membranes and hard dental surfaces. The heterogenous microbial species in the biofilms interact with the host environment and strongly communicate among each other through direct signals involving receptors and adhesins. An important characteristic of the biofilm is its inherent ability to extend protection from the invading competitive pathogenic microorganisms (Arweiler and Netuschil, 2016). The majority of the microbial organisms thriving in the oral cavity are commensal symbiotic organisms along with a few pathogenic strains in healthy individuals. However, when there is a clinical shift in the population of microbes and the normal organisms become pathogenic, biofilms exacerbate a variety of diseases and infections, resulting in irrevocable consequences.

Normal oral microbiota

The human oral microenvironment is replete with a diverse variety of bacteria, viruses, archaea species and protozoa, largely contributing to the well-being of the individual. Among this enriched population, bacteria dominate the habitat and flourish by multiplying rapidly. The mucous membrane lines the oral cavity, comprising both keratinized and non-keratinized epithelial cells. The cavity supports the growth of both anaerobes and aerobic bacteria, although predominantly facultative anaerobes or obligate anaerobic bacteria prevail in the mouth. The saliva, oral mucosa and gingival crevicular fluid (GCF) play a significant role in maintaining the ecology of the mouth (Xu *et al.*, 2015). The saliva proteins, glycoproteins and GCF sustain the development of several harmless bacteria. Gram-negative and Gram-positive bacteria are equally distributed in the microenvironment. A vast number of species of the bacterial genera *Streptococcus, Staphylococcus, Escherichia, Fusobacterium, Porphyromonas, Prevotella, Treponema, Neisseria, Haemophilis, Eubacterium, Lactobacillus, Lactococcus, Capnocytophaga, Eikenella, Leptotrichia* and *Propionibacterium* have been identified to contribute to the oral cavity microbiome (Kilian *et al.*, 2016). The neutral pH and optimum temperature of 35–36°C maintain the proliferation of the commensal microorganisms (Marsh *et al.*, 2009). The oral microfloral composition has been indicated to undergo variation along with the age of the individual. As a result, maintaining the potential oral microbial population is crucial for the ageing mouth (Illuzzi *et al.*, 2014). There is a fundamental interspecies relationship found in the community, leading to the pathogenesis of various diseases. A substantial number of species have been cultivated, but still many species are yet to be identified.

Oral cavity colonization

The indigenous microbiota colonizes the sterile oral cavity at birth as there is no oral microflora during fetal development. The maternal microbiota is transferred to the neonate through breastfeeding, subsequent to birth. The majority of the first formed microflora is composed of *Streptococcus salivarius, Candida albicans* and a few aerobic bacteria, which dominate until the appearance of teeth. There is a shift in the level of biodiversity over time until a point of homeostatic stability is reached. The bacterial community that populates the mouth after the eruption of the teeth includes various anaerobes such as *Streptococcus mutans, Streptococcus sanguis, Lactobacillus, Actinomyces, Veillonella* and *Spirochetes* (Sampaio-Maia *et al.*, 2014). The teeth provide an adhering surface to the bacteria for most of the organisms, while some are attached to the cheeks and gums. The tongue being a smooth surface with papillary structures has a copious population of Gram-positive actinomyces species and Gram-negative bacteroides species and peptostreptococcus species. Various external factors that influence the early microbial composition of the mouth include food, water consumed by the individual and horizontal bacterial transmission through communication with peer children (Avila *et al.*, 2009).

Prominent factors influencing the oral ecology

Hydrogen ion concentration

The optimal pH value of the saliva of a healthy individual lies between 6 and 7.9, whereas the pH value of the gingival crevice is about 6.9. Maintaining the ideal pH is crucial for protecting oneself from various consequences. Interestingly, neutral pH helps in developing and boosting the memory and cognitive ability as well as reducing the risk of stroke. When the pH value reduces below the expected optimal value, severe consequences begin to take a toll on the body functioning. The gums and teeth are severely affected by the acidic pH. This happens on repeated intake of acidic food, which includes those containing spices, herbs, animal protein, dairy foods, energy drinks and sugar-rich foods. The bacterial species act on the sugars such as glucose, fructose and sucrose and lead to the formation of dental carries. Surprisingly, many natural fruits provide lactic acid, which stimulates the formation of tooth decay. It is essential to maintain an alkaline pH by increasing the consumption of alkaline-rich foods such as vegetables like cucumbers, broccoli, sweet potatoes, spinach and avocados and fruits like mangos and lemons, which function to normalize the acidic environment of the mouth. Having said that, a strong alkaline pH (around 10–14) can also be detrimental for the body health. *Porphyromonas gingivalis*, a plaque-forming pathogenic microbe, is found to thrive at a pH of around 7.5 in the gingival crevices. A well-balanced pH is absolutely necessary to permit the growth of the commensal oral bacteria and inhibit the pathogenic strains (Fig. 6.1).

Temperature

The oral cavity temperature is a chief determinant of bacterial growth and as a result is crucial for maintenance of body health. An approximate value of 35–37°C is ideal for protecting the cavity from invading pathogens. Extremes of temperature can induce the multiplication of harmful bacterial strains. Considerable variation in temperature has been reported to occur between the teeth and oral mucosa.

Redox potential

Non-existence of oxygen supply is the prerequisite for the survival of obligate anaerobic bacteria in the mouth. Oxygen tension (pO_2) is an important factor for the growth of many bacterial species. Saliva is the source of oxygen supply in the mouth, with newly synthesized saliva containing an oxygen level of about 65 mmHg, which has the tendency to decrease to about 35–40 mmHg in about 60 seconds. The consumption of oxygen owing to the presence of certain pathogenic bacteria causes its reduction for breaking down nitrate and carbohydrate substrates. Studies have stated that the oxidation–reduction potential determines the pathogenesis of periodontal disease and dental plaque formation (Kenney and Ash *et al.*, 1969).

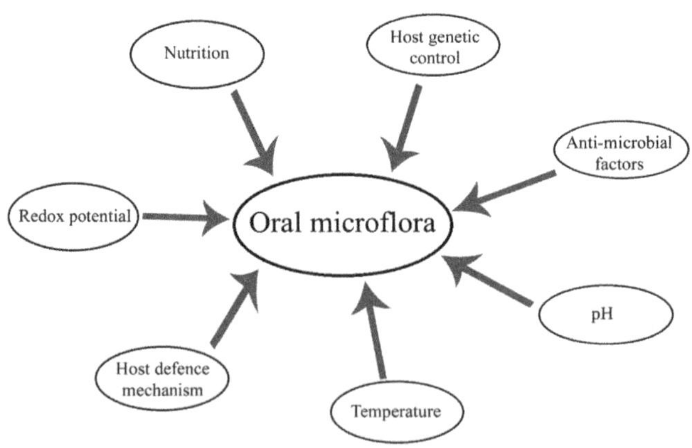

Figure 6.1 Factors affecting the oral microbiota.

Nutritional requirements

The endogenous and exogenous nutrient supply obtained from the diet shapes the oral community of bacterial species. Malnutrition is the initiating factor for the manifestation of poor oral health. Foods rich in complex carbohydrates and simple sugars yield substrates for the survival of many pathogenic bacteria. *Lactobacillus* strains and *Streptococcus mutans* are Gram-positive bacteria that flourish because of the abundant supply of carbohydrates and are implicated in the development of various diseases. An indirect association seems to exist between *Streptococci mutans* and the manifestation of haemorrhagic stroke. An abundant supply of micronutrients like calcium, phosphate and multivitamins is necessary for good oral hygiene. Deficiency of vitamins D, B_{12} and C and minerals can lead to the occurrence of gingivitis, tooth loss, swollen tongue and various other oral disorders and further resulting in systemic diseases. The salivary and GCF components enhance the stability of normal oral microflora. The abundant enzymes and proteins found in the saliva, in particular lysozyme, lactoferrin, statherin and lactoperoxidase, inhibit the development of pathogenic and cariogenic bacteria. Various other lipids and carbohydrates generated by the saliva, GCF and oral mucosa regulate the adherence of bacteria to the teeth and epithelial surfaces. The prevailing glycoproteins in the saliva contribute to the maintenance of the microenvironment and biofilm formation and function (van't Hof *et al.*, 2014).

Antimicrobial factors

Various antimicrobial agents present in oral health-care products as well as oral drugs and antibiotics can impair the structural and metabolic functions of the oral microbes. Numerous dental hygiene products including toothpaste, mouthwash and gels containing increased concentrations of antimicrobial agents such as fluoride, xylitol, iodine and chlorhexidine regulate the colonization of pathogenic microorganisms. Fluoride is a naturally existing mineral in water and trace amounts in food are known to have anticariogenic properties. It has been highly exploited by the health-care industry in the manufacture of various products and treatments. Fluoride has been investigated to alter the oral microbial composition by impairing the bacterial enzymes of various pathogenic strains. Enzymes such as enolases, sulphatases, F-ATPases and salivary peroxidases are inhibited, which destroys the plaque formation and acidification. Demineralization of the tooth is also highly reduced with exposure to fluoride. Various plaque-forming species including *Streptococcus mutans, Clostridium difficile, Lactobacillus casei* and *Fusobacterium nucleatum* are strongly inhibited by the use of fluoride-containing products. However, a high dosage of fluoride can have detrimental effects on the brain; as fluoride is neurotoxic it can lead to neurobehavioural changes, brain dysfunction and memory loss. Antibiotics are commonly known to alter the quantitative and qualitative composition of the host microbiota. Good commensal bacteria such as *S. sanguinis, Actinomyces naeslundii and Veillonella dispar.* present in the mouth are also affected by the administration of various drugs by the oral route. Studies have reported the incidence of various infections in the oral cavity owing to the use of prosthetic and orthodontic appliances. Palatal expanders, Forsus springs, Bionators, expanders and Turbo brackets have various modulatory and transitory effects on the oral microflora. They favour an increased colonization of pathogenic bacteria and fungal species such as *Streptococcus, Lactobacillus, Candida, Tannerella forsythia, Treponema denticola, Prevotella intermedia* and *Porphyromonas gingivalis*, resulting in tooth decalcification, plaque formation and inflammatory periodontal diseases.

Host defence mechanism

Cellular and humoral immune responses are activated against the invasive bacterial strains and are found to alter oral cavity composition. The anatomical factors and mucosal and epithelial barriers consisting of saliva and GCF present an attack against bacterial aggression in several periodontal diseases. Chronic inflammation induced by periodontitis stimulates the innate immune response. The subgingival cavity and epithelial barriers coordinate the first line of defence against the invading bacteria. The immune system components including macrophages, dendritic cells, immune T cells and B cells, known as antigen-presenting cells, and salivary antibodies and proteins are involved in the protection of the oral cavity (Walker *et al.*, 2004). Cytokines such as IL-2 and IL-8 are secreted against the invasive *Bacteroides gingivalis*, a harmful Gram-negative anaerobe, causing periodontal diseases. The flushing effects of the saliva protect the host oral microflora from various infective bacteria; similarly, the GCF works by modifying the microenvironment of the gingival crevice by destroying the microbial cells and provoking the immune cells to attack them. Innate immunity responses are triggered by sensing structures such as pattern recognition receptors (PRRs), which enable the immune cells to identify the bacterial lipopolysaccharides as well as to recognize the characteristic pathogen-associated molecular patterns to destroy them.

Host genetic control

The ecosystem of the oral cavity is predominantly predestined by the genetic make-up of the individual. The host genotype shapes the bacterial milieu in the mouth to a certain extent, However, they are not responsible for the formation of caries (Gomez *et al.*, 2017). The host oral microbiome has been established through the implementation of high-throughput sequencing techniques. Interestingly, it was found that identical twins share similar oral microflora as opposed to non-identical twins, who do not exhibit much similarity (Turnbaugh *et al.*, 2009).

Biofilm formation

A complex formation of intimately bonded bacterial species is found all over the oral cavity, primarily on the dental surfaces of the teeth as well as cheeks and gums. An expolymer matrix is formed as the base, which is a carbohydrate substance released by the bacteria. Biofilms are classified as uncalcified and calcified biofilms associated with etiopathology of various periodontal diseases. The normal oral microflora participate in the formation of biofilms, which act as a defensive mechanism against the invasion of competitive bacteria as well as providing protection and maintenance of the microenvironment by facilitating the uptake of nutrients and regulation of the redox potential. The bacterial species strongly interact with each other through the biofilm formation and are responsible for transfer of genes among them. Conversely, biofilms can become pathogenic by the colonization of invasive antibiotic-resistant bacteria and can initiate the development of various systemic diseases. The biofilm undergoes expansion on maturation and the developing pathogenic bacteria complex can accelerate the manifestation of dental plaques, gingivitis and periodontitis (Costerton *et al.*, 1999). The oral biofilm formation in patients diagnosed with periodontitis has been exhibited to result in various adverse events, especially brain-related infirmities.

The mouth–brain connection

Neurology and dentistry have unfamiliar links and increasingly many dental pathogens are promoting a wide spectrum of neurological malfunctions such as stroke, Alzheimer's disease, brain abscess, epilepsy, dementia, Parkinson's disease, brain aneurysm and schizophrenia. Infections in the oral cavity including dental plaque formation, root canals and periodontal diseases have the ability to increase the risk of attaining neurological diseases. This is primarily because of the pervasive formation of blood clots within the blood vessel and spread of embolisms in the jugular vein. The pathogenic bacteria in these blood clots spread rapidly to the cranial vein and to the lateral sellar compartment, which is bordered by the temporal bone of the skull. The oral pathogens cause detrimental effects to the brain once they have gained access to the bloodstream.

Stroke

One of the most dreaded diseases in the present age is stroke affecting the brain, which is the major control system of the body. The arteries carrying blood to the brain cells are obstructed, impairing the functioning of the brain. If left neglected, this can lead to permanent brain damage as well as death. The approximate estimate of the global annual incidence of stroke is about 15 million people globally. There are two different kinds of stroke depending on the location of the forming blood clots. One of the major factors contributing to the occurrence of haemorrhagic stroke is persistent high blood pressure. The disease causes immense distress, impacting vision and memory potential, and is the chief causal factor of long-term disability, spasticity, weakness and fatigue. Various determinants of stroke include high blood pressure, obesity and lifestyle habits; in addition, more recently, oral health has been identified to be a key factor. Studies have unveiled the fatal association between oral pathogens and development of a stroke. A particular bacterial strain, namely *Streptococcus mutans*, universally known for dental plaque formation, has now been found to initiate the rupture of the arteries supplying blood to the brain. The oral bacteria gain entry into the blood by causing bacteraemia and then further attack the brain leading to vascular injuries and brain dysfunction. This understanding can further assist in developing oral probiotics to counteract the presence of harmful oral pathogens as well as stressing the need for proper oral health care (Sfyroeras *et al.*, 2012).

Alzheimer's disease

Various bacterial strains are known to colonize the oral cavity, predisposing people to various infections such as periodontitis and formation of dental caries. However, additionally, they have been found to increase the risk of cognitive decline in people affected with Alzheimer's disease. Alzheimer's disease is a widely known neurodegenerative disease primarily affecting the neurons in the hippocampal region of the brain. Currently, worldwide about 50 million people have been diagnosed with Alzheimer's disease and this is expected to double, with nearly 75 million people affected by 2030. The hallmark characteristic of Alzheimer's disease is formation of neurofibrillary tangles and beta-amyloid plaques within the brain cells, as well as severe inflammation of the brain. Deterioration of memory and reasoning ability, poor judgement and withdrawal from social activities are the predominant manifestations of the disease. The accumulated beta-amyloid protein in the fatty membrane surrounding the nerve cells and formation of twisted nerve fibres composed of tau protein leads to the development of the disease. The origin of the disease can be due to either the transmission of genetic traits or various environmental factors. Although much has been researched about the role of periodontal pathogens and the progress of Alzheimer's disease, the bacterial species resulting in biofilm formation are known to initiate a host immune response, provoking the release of various inflammatory cytokines. The biofilm, largely composed of Gram-negative bacteria, can enter the systemic circulation and the bacterial pathogens have the ability to disseminate to various anatomical sites in the body. The majority of bacterial strains responsible for the development of periodontitis include *Porphyromonas gingivalis, Fusobacterium nucleatum, Tannerella forsythensis* and *Treponema denticola*. These microorganisms gain access through neuronal or peripheral nerve pathways by various mechanisms and stimulate the production of interleukins, tumour necrosis factor-alpha (TNF-a) and chemokines (Sochocka *et al.*, 2017). The cytokines have the ability to compromise the integrity of the blood–brain barrier and result in serious consequences, impairing the interaction between the glial cells and neurons. *Porphyromonas gingivalis* is a non-motile Gram-negative anaerobe that can destroy the periodontal tissues and induce collagen degradation. It is one of the predominant species promoting periodontal disease. Its virulence factors elevate the level of the antibodies in the blood. Several reports have confirmed the presence of *Porphyromonas gingivalis, Chlamydophila pneumoniae* and *Aggregatibacter actinomycetemcomitans* in various regions of the brain, causing several brain infections (Olsen and Singhrao, 2015).

Multiple sclerosis

This is an autoimmune disease impacting the central nervous system whereby it obstructs the transmission of electric signals carrying information from the brain to the body. It disrupts the myelin sheath surrounding the brain and spinal cord nerve cells and exacerbates the immune system response. The factors contributing to the development of the disease are manifold, including immunological, environmental, genetic and various other infections. Approximately around 2,500,000 people are diagnosed with multiple sclerosis worldwide. It is a devastating neurological disease causing various symptoms such as loss of balance, numbness, lack of coordination and difficulties with speech and walking. Common bacterial pathogens have the capability to have effects beyond the oral cavity; collectively various study reports have established their link with the generation of several neurological diseases. The oral pathogens lead to the formation of clots in the blood vessel supplying the brain, which results in destruction of the myelin sheath of the nerve axons in the brain stem. Various bacterial byproducts are responsible for triggering the severity of the disease. The well-known periodontal pathogen *Porphyromas gingivalis* has been indicated to heighten the immune response by the production of a lipid substance such as phosphorylated dihydroceramides (DHCs), which have the ability to activate the immune cells to generate an inflammatory reaction (Elemek and Almas, 2013). Researchers have found that lipid phosphorylated DHCs may have the potential to be classified as a marker of multiple sclerosis and should be investigated as a new target for therapeutic interventions.

Brain abscess

Various microorganisms have the ability to affect the brain, causing inflammatory damage to the brain parenchyma, swelling and accumulation of dead cells in the brain, thereby resulting in an abscess. The causative agents include several Gram-negative and Gram-positive anaerobic bacteria as well as a few microaerophilic cocci, namely *Staphylococcus aureus, Bacteroides, Prevotella* and *Fusobacterium* species, Enterobacteriaceae, *Pseudomonas* species,

Haemophillus influenzae, Streptococcus pneumoniae and *Neisseria meningitidis,* of which some are habitually present in the oral cavity (Patel *et al.*, 2014). These bacterial species can reach the brain by directly spreading through an oral infection, during a dental treatment procedure, or by haematogenous spread through facial, angular or ophthalmic veins and via local lymphatics. The pathogens bind to a specific Toll-like receptor, resulting in enhanced activation of astrocytes and microglia stimulating an increased production of pro-inflammatory cytokines, which permeate the brain. In particular, *Fusobacterium nucleatum* has been extensively studied for its association with the development of multiple brain abscesses. This Gram-negative anaerobe is a constituent member of the oral cavity, which under unfavourable conditions has been vividly expressed in the pathogenesis of chronic periodontal diseases and reversible gingivitis. It extensively consumes dietary carbohydrates as its energy source in the formation of dental plaques (Avila-Campos and Nakano, 2006). A wide array of environmental factors is fundamental in provoking this oral commensal organism in stimulating the host immune response. Various cytokines including IL-1, IL-6 and TNF-a are secreted and are involved in the inflammation-mediated reactions. Treatment of brain abscess generally involves a collaborative effort of medication and surgery, depending on the severity and size of the abscess. It is highly essential to eliminate the causative pathogen in order to attain successful therapy.

Brain aneurysm

The weakness and bulge formation in the cranial blood vessels such as the vertebrobasilar and carotid arteries are also found inside the brain result in the origination of brain aneurysm, also referred to as intracranial or cerebral aneurysm. Researchers have found that a high percentage of oral pathogens are significantly involved in the pathogenesis of this medical condition. A weakened blood vessel can be exceedingly dangerous; if it leads to a rupture it can permanently damage the brain. *Aggregatibacter actinomycetemcomitans, Fusobacterium nucleatum* and *Treponema denticola* were identified as potential risk factors for the incidence of intracranial aneurysm disease and the rupture of brain aneurysms.

Parkinson's disease

Parkinson's disease is quite a well-known multifactorial degenerative disease that injures the neurons that are responsible for the production of the neurotransmitter dopamine in the brain. The substantia nigra is an important midbrain structure associated with the dopaminergic neurons. Dopamine plays an important role in regulating motor control and movement, behaviour, emotional responses, cognitive potential and learning ability. Severe oral health complications increasingly promote the progress of the disease as well as appearing as symptoms of this condition. Scientific evidence has pointed out the variation in the composition of the oral flora in patients with Parkinson's disease. Further research into the oral microbiome of patients with the disease is necessary to identify a potential biomarker and therapeutic strategy for alleviating oral infections.

Management of oral health

With the rise of numerous harmful bacteria colonizing the oral cavity and predisposing to a spectrum of neurological diseases, it is mandatory to counteract the inhibitory clinical effects of these pathogenic strains. The field of prebiotics and probiotics has gained overwhelming importance in sustaining the symbiotic population of microbes in the host system. Notable bacterial species exploited for their beneficial role in conferring health include many species of lactobacilli, lactococci and streptococci. They exhibit manifold therapeutic effects, such as inhibiting oral diseases, which points to periodontitis, inflammation of gums and oral candidiasis, as well as being influential in hindering the oral manifestations of systemic diseases leading to neurological and psychological aberrations (Meurman and Stamatova, 2007). Novel probiotics are being developed with the objective to combat highly virulent pathogenic strains such as *Actinomyces, Porphyromonas gingivalis, Prevotella intermedia* and *Streptococcus mutans* in order to maintain healthy teeth and gums and also decrease the inflammation-associated response in the mouth (Kumar *et al.*, 2017). Increasingly, various *Lactobacillus* species have been clinically investigated for their protective effects against the invading microbial pathogens. *L. bulgaricus, L. rhamnosus, L. reuteri, L. acidophilus* and *L. casei* are known to be extensively utilized for their bactericidal and bacteriostatic properties. Their mode of action is through the production of organic acids, specifically lactic and acetic acid, and particular metabolites such as hydrogen peroxide and diacetyl, besides important peptides and proteins that have inherent antimicrobial activity (Stamatova *et al.*, 2007). Porphyromonas gingivalis is the principal causative agent of periodontitis infection, which is the leading cause of dissemination to the brain and spinal cord. The pathogen exerts its unfavourable effects with the aid of several virulence factors such as lipopolysaccharide (LPS), haemagglutinins, capsules and predominantly fimbriae, which are present as cell surface components. This bacterial strain is composed of a fimA gene, which is the essential determinant of the filamentous component and has six distinguished genotypes, each of which are highly virulent. A clinical investigation has been conducted to study the therapeutic effects of *Lactobacillus* strain L1 against the distinct *fimA* genotypes of the microbe (Matsuoka and Koga, 2014). It is believed that the epithelial lining of the oral cavity plays an important role in defending the host from invading microbial communities. Probiotics are known to exert immunomodulatory signals, which are required for the stimulation of host immunity. Experimental studies have evaluated the significance of probiotics on the inflammatory characteristics of oral epithelial cells, along with the pathogenic *Porphyromonas gingivalis* (Stamatova *et al.*, 2010).

Summary and conclusion

Oral health problems such as dental cavities are often neglected by individuals and not given much medical attention unless they result in agonizing pain. Little did people know about the importance of oral hygiene in association with maintenance of brain health? Recent study reports have helped in revealing the relationship between oral microflora and the progression of diverse neurological disorders. This chapter outlines that an understanding of the oral microbiota is as equally important as an understanding of gut bacterial data. Considerable work has been published in the field of the gut–brain axis and its related consequences due to dysbiosis. However, not much has been mentioned regarding the effects of oral pathogens on the brain. Emphasizing the significance of the normal oral microflora and its imbalance resulting in the development of unanticipated consequences has precautioned neurologists, scientists, physicians, and ultimately common individuals to ensure proper oral health-care regime. This revelation will encourage further research in the field of oral microbiome to identify and develop novel biomarkers and therapeutic interventions.

Acknowledgements

The authors are thankful to the Chettinad Academy of Research and Education (CARE), Chennai, India, for the research facilities and support.

Funding

This study was partially financially supported by grants from the Chettinad Hospital and Research Institute (CHRI), Chettinad Academy of Research and Education (CARE), Chennai, India.

Conflicts of interests

No potential conflict of interest was reported by the authors.

Author contributions

SVP, SS GJ and AB wrote the manuscript. SP, RM and FM designed and revised the manuscript.

References

Aas, J.A., Paster, B.J., Stokes, L.N., Olsen, I., and Dewhirst, F.E. (2005). Defining the normal bacterial flora of the oral cavity. J. Clin. Microbiol. *43*, 5721–5732.

Arweiler, N.B., and Netuschil, L. (2016). The oral microbiota. In Microbiota of the Human Body: Implications in Health and Disease, Schwiertz, A., ed., Adv. Exp. Med. Biol. *902*, pp. 45–60. https://doi.org/10.1007/978-3-319-31248-4_4.

Avila, M., Ojcius, D.M., and Yilmaz, Ö. (2009). The oral microbiota: living with a permanent guest. DNA Cell Biol. *28*, 405–411. https://doi.org/10.1089/dna.2009.0874.

Avila-Campos, M.J., and Nakano, V. (2006). Pathogenicity of *Fusobacterium nucleatum*: General aspects of its virulence. Int. J. Probiotics Prebiotics *1*, 105–112.

Costerton, J.W., Stewart, P.S., and Greenberg, E.P. (1999). Bacterial biofilms: A common cause of persistent infections. Science *284*, 1318–1322.

Elemek, E., and Almas, K. (2013). Multiple sclerosis and oral health: An update. N. Y. State Dent. J. *79*, 16–21.

Dewhirst, F.E., Chen, T., Izard, J., Paster, B.J., Tanner, A.C., Yu, W.H., Lakshmanan, A., and Wade, W.G. (2010). The human oral microbiome. J. Bacteriol. *192*, 5002–5017. https://doi.org/10.1128/JB.00542-10.

Gomez, A., Espinoza, J.L., Harkins, D.M., Leong, P., Saffery, R., Bockmann, M., Torralba, M., Kuelbs, C., Kodukula, R., Inman, J., *et al.* (2017). Host genetic control of the oral microbiome in health and disease. Cell Host Microbe *22*, 269–278.e3.

Illuzzi, N., Galli, R., Kushugulova, A., Zhumadilov, Z., Licciardello, O., and Marotta, F. (2014). Expanding the Metchnikoff postulate: Oral health is crucial in a successful global aging management strategy. Rejuvenation Res. *17*, 172–175. https://doi.org/10.1089/rej.2013.1493.

Kenney, E.B., and Ash, M.M. (1969). Oxidation reduction potential of developing plaque, periodontal pockets and gingival sulci. J. Periodontol. *40*, 630–633. https://doi.org/10.1902/jop.1969.40.11.630.

Kilian, M., Chapple, I.L., Hannig, M., Marsh, P.D., Meuric, V., Pedersen, A.M., Tonetti, M.S., Wade, W.G., and Zaura, E. (2016). The oral microbiome – an update for oral healthcare professionals. Br. Dent. J. *221*, 657–666. https://doi.org/10.1038/sj.bdj.2016.865.

Kumar, N., Marotta, F., Dhewa, T., Mishra, V., Kumar, V., and Bharadwaj, A. (2017). Management of oral health through novel probiotics: A review. Int. J. Probiotics Prebiotics *12*, 109–114.

Meurman, J.H., and Stamatova, I. (2007). Probiotic applications in the oral cavity. Int. J. Probiotics Prebiotics 2, 1–8.

Olsen, I., and Singhrao, S.K. (2015). Can oral infection be a risk factor for Alzheimer's disease? J. Oral Microbiol. 7, 29143. https://doi.org/10.3402/jom.v7.29143.

Marsh, P.D., and Martin, M.V. (2009). The mouth as a microbial habitat. In Oral Microbiology, 5th ed., Marsh, P.D., and Martin, M.V. (Churchill Livingstone Elsevier, Edinburgh, London, New York, Oxford), pp. 8–23.

Matsuoka, T., and Koga, Y. (2014). The differences in the sensitivity of the *Porphyromonas gingivalis* fimA genotypes to a probiotic *Lactobacillus* strain, LS1, in the oral cavity of patients: Double blind clinical trial. Int. J. Probiotics Prebiotics 9, 135–140.

Patel, K., and Clifford, D.B. (2014). Bacterial brain abscess. Neurohospitalist *4*, 196–204. https://doi.org/10.1177/1941874414540684.

Sfyroeras, G.S., Roussas, N., Saleptsis, V.G., Argyriou, C., and Giannoukas, A.D. (2012). Association between periodontal disease and stroke. J. Vasc. Surg. *55*, 1178–1184. https://doi.org/10.1016/j.jvs.2011.10.008.

Sampaio-Maia, B., Caldas, I.M., Pereira, M.L., Pérez-Mongiovi, D., and Araujo, R. (2016). The oral microbiome in health and its implication in oral and systemic diseases. Adv. Appl. Microbiol. *97*, 171–210.

Sochocka, M., Zwolińska, K., and Leszek, J. (2017). The infectious etiology of Alzheimer's disease. Curr. Neuropharmacol. *15*, 996–1009. https://doi.org/10.2174/1570159X15666170313122937.

Stamatova, I., Kari, K., and Meurman, J.H. (2007). *In vitro* evaluation of antimicrobial activity of putative probiotic Lactobacilli against oral pathogens. Int. J. Probiotics Prebiotics 2, 225–232.

Stamatova, I., Kari, K., and Meurman, J.H. (2010). Probiotic challenge of oral epithelial cells *in vitro*. Int. J. Probiotics Prebiotics 5, 13–20.

Turnbaugh, P.J., Ley, R.E., Hamady, M., Fraser-Liggett, C.M., Knight, R., and Gordon, J.I. (2007). The Human Microbiome Project. Nature *449*, 804–810. https://doi.org/10.1038/nature06244.

van 't Hof, W., Veerman, E.C., Nieuw Amerongen, A.V., and Ligtenberg, A.J. (2014). Antimicrobial defense systems in saliva. Monogr. Oral Sci. *24*, 40–51. https://doi.org/10.1159/000358783.

Walker, D.M. (2004). Oral mucosal immunology: An overview. Ann. Acad. Med. Singap. *33* (Suppl. 4), 27–30.

Xu, X., He, J., Xue, J., Wang, Y., Li, K., Zhang, K., Guo, Q., Liu, X., Zhou, Y., Cheng, L., *et al.* (2015). Oral cavity contains distinct niches with dynamic microbial communities. Environ. Microbiol. *17*, 699–710. https://doi.org/10.1111/1462-2920.12502

The Influence of Gut Microbiota's Composition on the Carcinogenesis of Gastrointestinal Tract Cancers

7

Vo Phuoc Tuan,[1,2] Boldbaatar Gantuya,[1,3]
Takashi Matsumoto[1] and Yoshio Yamaoka[1,4*]

[1]Department of Environmental and Preventive Medicine, Faculty of Medicine, Oita University, Yufu, Japan.
[2]Department of Endoscopy, Cho Ray Hospital, Ho Chi Minh, Vietnam.
[3]Department of Gastroenterology, Mongolian National University of Medical Sciences, Ulaanbaatar, Mongolia.
[4]Department of Medicine, Gastroenterology and Hepatology Section, Baylor College of Medicine, Houston, TX, USA.

*Correspondence: yyamaoka@oita-u.ac.jp

https://doi.org/10.21775/9781910190937.07

Abstract

Interest in the role of microbiota in human health has increased over the last several decades. Current advanced techniques have made it possible to study the compositions and the functions of microbial communities in humans, especially in the gastrointestinal tract where the majority of bacteria reside. The disturbances of these communities, so-called 'dysbiosis', have been implicated to be associated with gastrointestinal tract cancers; however, the patterns or the role of the dysbiosis are still not yet fully understood. Recently, several microbiota-based therapeutic approaches have been developed; however, the ability to apply them as well as the efficacy of these treatments in clinical practice remain controversial. From this context, this chapter aims to summarize the recent findings, focusing on the bacterial dysbiosis in the gastrointestinal tract, particularly in the oesophagus, stomach and colon, and we discuss their roles in carcinogenesis. In addition, we describe the potential application of therapeutic approaches targeting microbiota and discuss the future trend of microbiota studies.

Introduction

Since the first bacterium was observed by Antoni van Leeuwenhoek in 1680s, a theory in which many diseases are caused by microorganisms (so-called germ theory) was officially accepted based on Koch's postulates. This theory opened a golden era for bacteriology and boosted the development of infectious disease research. Then, many causative microbial agents were discovered; however, these pathogens were mainly identified based on microscopy and culture-based methods. Recently, a revolution owing to the development of advanced techniques including culture-independent methods (i.e. high-throughput DNA sequencing) and bioinformatics have allowed researchers to focus on the contents and the functions of microbial communities, not only single specific microorganism (Schwabe and Jobin, 2013).

The microbiota is defined as the pool of entire commensal microorganisms (comprising bacteria, archaea, viruses and unicellular eukaryotes) colonizing a particular location of our body and is mostly found at the surface between our body and the outside world (Turnbaugh *et al.*, 2007). Human microbiota was known to play a key role on many important functions that lead to nutritional, metabolic, immunological and developmental benefits. However, the microbes are also important modulators of host phenotypes; therefore, the disturbance of human microbiota (so-called dysbiosis) was proven to be associated with an array of disorders or diseases in different body systems such as the gastrointestinal (GI), cardiovascular, respiratory, metabolic and autoimmune systems. Microbiome studies have revealed the inter-individual variations, heterogeneity of bacterial communities, functional redundancy along and across the different organs of human body and the need to distinguish cause from effects in states of dysbiosis.

With the most densely populated microbial communities, gut or GI microbiota appears to become one of the most crucial human microbial community and it represents a complex microbial ecosystem in a hollow tube from the oral cavity to the anus canal. The gut microbiota confers many benefits to the host through physiological functions ranging from strengthening gut integrity or shaping the intestinal epithelium to harvesting energy, protecting against pathogens and regulating host immunity (Thursby and Juge, 2017). Because the digestive tract is an opened hole to expose with the outside world and is an area colonized by the most abundant bacteria, especially a major part of 10^{13} bacteria in the large intestine, the frequent and continuous exposure of the gut epithelial cell barrier to these microbiota is inevitable and poses a huge threat to gut integrity (Sender *et al.*, 2016). Indeed, altered gut bacterial composition has been known to be associated with the pathogenesis of various clinical presentations that range from benign, inflammatory diseases (e.g. inflammatory bowel disease, irritable bowel syndrome) to malignant, severe diseases such as oesophageal cancer (EC), gastric cancer (GC) or colorectal cancer (CRC) (Nardone *et al.*, 2017).

Statistical data revealed that the GI cancers are the most important among all cancers, accounting for 25% of the incidence of new cases and 40% of cancer-related mortality worldwide; infectious agents were estimated to be attributed to 2.2 million new cancer cases (15.4%) (Plummer *et al.*, 2016; Torre *et al.*, 2015). The clinical outcomes of altered gut microbiota vary widely, not only from intestinal diseases but also from extra-intestinal diseases such as neurodevelopmental disorders, atherosclerosis and allergic asthma, with differences on the severe level ranging from benign, inflammatory to malignant lesions (Schroeder and Bäckhed, 2016). The other presentations including oral and neurological diseases will be presented in the other chapters of this book. Because of the rarity of the incidence of small intestine cancer compared with the other GI tract, the present chapter aims to summarize the current knowledge, given extensive research and data availability, about the composition of the human gut microbiota and its impact on the common GI tract cancers: EC, GC and CRC. In this chapter, we focus on bacteria as the most abundant and well-studied component of the gut microbiota. Furthermore, a recent novel approach, the so-called dual therapeutic strategy (a combination of immunotherapy and microbiota-targeted approaches), was applied in dysbiotic microbiota-related diseases. This information underlies and paves the way for future studies about the new, putative pathogens in specific malignant diseases of the GI tract as well as the novel screening of or treatment strategies for these diseases.

Gut microbiota composition

Factors shaping the gut microbiota

Mode of delivery

Even though it is widely believed that the gut is colonized by microbes immediately after birth, there is emerging evidence that the infant gut could be colonized by organisms even *in utero* (Dominguez-Bello *et al.*, 2010). Nevertheless, it is now clear that the first microbiota profile is largely shaped by the mode of delivery. The intestines of infants born vaginally are initially colonized by organisms from the maternal vagina, which is best exemplified by the organisms from the genera *Lactobacillus* and *Prevotella*. On the contrary, in caesarean delivery mostly the maternal skin flora colonizes the infant's intestine, as exemplified by the dominance of *Streptococcus*, *Corynebacterium* and *Propionibacterium* (Dominguez-Bello *et al.*, 2010; Mackie *et al.*, 1999).

Age

The initial milieu of the infant's gut microbiota after primary inoculation appears to be unstable and devoid of diversity; however, with time it stabilizes, diversifies and acquires 40–60% similarity to the adult microbiota by the age of three years (Yatsunenko *et al.*, 2012). On the contrary, studies have also shown that young children and adolescents could demonstrate significant differences in proportions of *Bacteroides* and *Bifdobacterium* compared with adults (Agans *et al.*, 2011; Ringel-Kulka *et al.*, 2013). The gut microbiota seem to rest in a stable state from the third to the seventh decades of life, even though proportions of *Bifidobacteria*, Firmicutes and *Faecalibacterium prausnitzii* tend to gradually decrease, with an increase in *E. coli*, Proteobacteria and *Staphylococcus* (Biagi *et al.*, 2010; Claesson *et al.*, 2011; Mariat *et al.*, 2009).

Diet

The earliest presentation about the impact of diet on the gut microbiota is the infant diet, including breast milk and formula feeds. It is important to understand the effect of breast milk and formula feeds on the gut microbiota as there has been an increasing trend of moving away from breastfeeding by modern mothers. The breastfed infants showed a significantly higher abundance of *Bifidobacterium* spp. and lower abundance of *Bacteroides* spp. and *Clostridium* spp. than formula-fed infants (Harmsen *et al.*, 2000; Stark *et al.*, 1982; Yoshioka *et al.*, 1983; Zivkovic *et al.*, 2011).

Diet continues to be the most important determinant in shaping the composition, diversity and richness even throughout adulthood. In general, intake of a diet rich in fruits, vegetables and fibre is associated with a higher richness and diversity of the gut microbiota. Individuals consuming this kind of a diet have a higher abundance of the insoluble carbohydrate-metabolizing organisms of the Firmicutes phylum such as *Ruminococcus bromii*, *Roseburia* and *Eubacterium rectale* (Walker *et al.*, 2011). It was recently shown that four days of administration of an animal-based diet resulted in a decrease in the abundance of Firmicutes and an increase in that of bile-tolerant organisms such as *Alistipes* spp. and *Bacteroides* spp. from the phylum Bacteroidetes and *Bilophila* spp. from the phylum Proteobacteria. This indicates that even very short dietary manipulations can have a substantial impact on the gut microbiota (David *et al.*, 2014).

Use of antibiotics

Because of bactericidal and bacteriostatic activities against pathogens, antibiotics became a prominent subject for researchers to study their effects on gut bacterial ecology. Many scientific evidences showed that the use of antibiotics has several short- and long-term impacts on the ecology of the normal gut microbiota. One of the major properties of the healthy gut microbiota against pathogen is the capability to cause competitive exclusion. Major changes in the gut microbiota in response to the use of antibiotics include the depletion of taxonomic diversity and persistence of the changes in a substantial proportion of individuals. For example, Jakobsson *et al.* (2010) showed that the diversity of Actinobacteria dramatically reduced after a course of *Helicobacter pylori* eradication therapy with traditional triple therapy (clarithromycin, amoxicillin and proton-pump inhibitor) and the effects persisted for up to 4 years in some cases.

Normal gut microbiota composition in the GI tract

The number of bacteria declines from the oral cavity to the stomach and then markedly increases from the small intestines to reach a maximum number in the colorectum (Lam *et al.*, 2017). The healthy GI microbiota varies considerably among individuals and body sites, but the common gut microbial species predominantly belong to four major phyla (Firmicutes, Bacteroidetes, Proteobacteria and Actinobacteria) (Fig. 7.1) (Human Microbiome Project Consortium, 2012). The phylum Firmicutes mainly comprises the genera *Streptococcus* and *Bacillus*; the phylum Bacteroidetes comprises the *Prevotella* species; the phylum Proteobacteria comprises the genera *Helicobacter*, *Haemophilus*, *Actinobacillus* and *Neisseria*; and the phylum Actinobacteria comprises the *Rothia*, *Actinomyces* and *Micrococcus* species. Because of the various distribution of bacterial load/type in different sites of the digestive tract, the data about normal gut microbiota composition based on individual organs are presented in the following sections.

Oesophagus

Because the oesophagus is considered as a transitory place that transports food from the mouth to the stomach, it was hypothesized that the oesophagus was colonized by only transient bacteria due to its rapid peristalsis and the impact of the lower oesophageal sphincter. Initial studies using culture-based methods showed the main components of oesophageal microbiota were *Streptococcus* spp., with *Streptococcus viridans* dominating (Gagliardi *et al.*, 1998; Norder Grusell *et al.*, 2013; Pajecki *et al.*, 2002; Zilberstein *et al.*, 2007); however, these reports underestimated the complexity of the oesophageal bacterial ecosystem (Table 7.1). The first study, conducted by Pei *et al.* (2004) using a culture-independent method

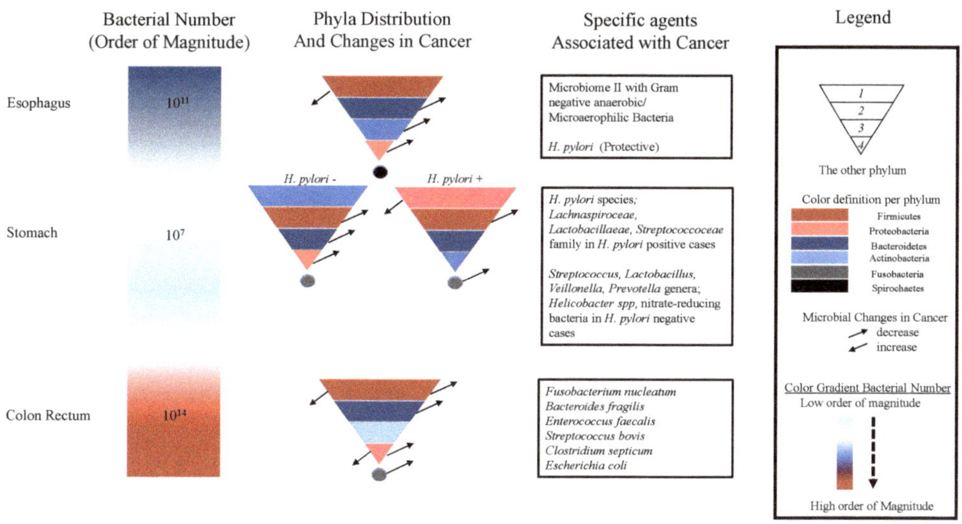

Figure 7.1 The human gastrointestinal microbiome composition in health and cancer.

Table 7.1 The composition of normal oesophageal microbiota

Articles	Patients	Sample type	Methods	Results
Gagliardi *et al*. (1998)	30 healthy volunteers	Culture	Luminal wash	*Streptococcus viridans* was the predominant species
Pajecki *et al*. (2002)	10 normal oesophageal patients	Culture	Luminal wash	*Streptococcus* genus was the predominant species
Pei *et al*. (2004)	4 normal oesophageal patients	16S rDNA PCR	Biopsy specimens	95 species and 41 genera clustering into 6 phyla (Firmicutes, Bacteroides, Actinobacteria, Proteobacteria, Fusobacteria and TM7); *Streptococcus*, *Prevotella* and *Veillonella* were the predominant genera
Pei *et al*. (2005)	9 normal oesophageal patients	16S rDNA PCR	Biopsy specimens	4 phyla (Bacteroides, Firmicutes, Proteobacteria and Actinobacteria)
Zilberstein *et al*. (2007)	10 healthy volunteers	Culture	Aspirate	*Streptococcus* spp. (40%), *Staphylococcus* spp. (20%), *Corynebacterium* spp. (10%), *Lactobacillus* spp. (10%), and *Peptococcus* spp. (10%)
Yang *et al*. (2009)	12 normal oesophageal patients	16S rRNA gene clone	Biopsy	*Streptococcus* genus was the predominant species
Fillon *et al*. (2012)	15 children with normal histology	16S rDNA gene sequencing	Oesophageal string test; biopsy	5 phyla (Actinobacteria, Bacteroidetes, Firmicutes, Fusobacteria and Proteobacteria); 31 genera with the main components being *Streptococcus*, *Prevotella* and *Veillonella*

Table 7.1 Continued

Articles	Patients	Sample type	Methods	Results
Liu *et al.* (2013)	6 normal oesophageal patients	16S rDNA gene sequencing	Biopsy specimens	4 phyla (Proteobacteria, Firmicutes, Bacteroidetes and Actinobacteria);11 genera with the main components being *Streptococcus* and *Klebsiella*
Norder Grusell *et al.* (2013)	40 normal oesophageal patients	Culture	Brushing; biopsy specimens	23 species with *Streptococcus viridans* being the most common, followed by *Fusobacterium* spp., *Neisseria* spp., *Haemophilus* spp. and *Prevotella* spp.

in order to characterize the oesophageal flora, revealed a higher resolution of the microbial community with the presence of 95 species and 41 genera clustering into six phyla with the predominance of Firmicutes, followed by Actinobacteria, Proteobacteria, Bacteroidetes, Fusobacteria and TM7. With similar findings, subsequent studies also confirmed that major components of the core microbiota of the healthy oesophageal mucosa belong to four phyla (Firmicutes, Bacteroidetes, Actinobacteria and Proteobacteria) (Fillon *et al.*, 2012; Liu *et al.*, 2013; Pei *et al.*, 2005; Yang *et al.*, 2009). Interestingly, at genus level the most frequent core microbiota was *Streptococcus*, irrespective of methodological approaches (culture dependent vs. culture independent) and different analysed specimens (aspirate fluid vs. biopsy specimen) (Fillon *et al.*, 2012; Gagliardi *et al.*, 1998; Liu *et al.*, 2013; Norder Grusell *et al.*, 2013; Pajecki *et al.*, 2002; Pei *et al.*, 2004; Yang *et al.*, 2009; Zilberstein *et al.*, 2007).

Despite the data about the composition of the normal oesophagus flora being scarce, in summary, the oesophageal flora was characterized by the predominance of Firmicutes, followed by Bacteroidetes, Actinobacteria and Proteobacteria at phylum level and *Streptococcus* spp. at the genus level.

Stomach

Historically, the stomach was long believed to be 'a sterile organ' because of the low pH of gastric acid (Dias-Jacome *et al.*, 2016). The current studies showed that many bacteria are able to thrive in the stomach; however, the estimation of 10^7 bacterial abundances in the stomach presents a lower colonization than in other GI compartments, because of the relatively hostile condition of the gastric environment (Lancet, 1981). *H. pylori* infection is the most common infectious disease in the stomach, with more than half of the world's population infected. It was recognized that the colonization of this bacterium is one of the most important factors to shape the microbial ecosystem in this organ. However, the fact that most of *H. pylori*-infected people are asymptomatic, suggesting that the bacterium could play some positive role in the evolution of human beings or it could be only a commensal. Until now, the definition of a normal gastric flora is still controversial; however, in this chapter, we considered a healthy control or gastric flora as the *H. pylori*-negative patients with normal endoscopic examination or healthy volunteers without *H. pylori* infection.

To date, only a few studies focused on describing the gastric microbiota of *H. pylori*-negative healthy people without any GI tract symptoms. There were only two studies conducted on healthy volunteers. First, Zilberstein *et al.* (2007), by using culture from fluid aspiration, displayed that the most abundant bacteria in the stomach belonged to five phyla (Firmicutes, Proteobacteria, Actinobacteria, Fusobacteria and Bacteroidetes) and 18 genera, with the main components being *Veillonella* spp., followed by *Lactobacillus* spp. and *Clostridium* spp. Subsequently, Andersson *et al.* (2008), with higher resolution by using 16S rRNA pyrosequencing, indicated that the gastric flora consisted of 262 phylotypes belonging to 13 phyla. The phylotypes were divided into two groups; one was similar to bacteria in the throat (belonging to *Streptococcus*, *Actinomyces*, *Prevotella* and *Gemella* genera), representing swallowed microorganisms, and the other was found commonly in the stomach but not in the throat (mostly belonging to the Proteobacteria phylum).

Other remaining studies considered the gastric flora derived from *H. pylori*-negative patients with normal endoscopic examination. The first profiling of gastric microbiota other than *H. pylori* reported by Monstein *et al.* (2000) revealed *Enterococcus*, *Pseudomonas*, *Streptococcus*, *Staphylococcus* and *Stomatococcus* as the most abundance genera and they belonged to three major phyla (Proteobacteria, Firmicutes and Actinobacteria). Despite most bacteria inhabiting the oral cavity and respiratory tract, the presence of *Pseudomonas* spp. other than *Pseudomonas aeruginosa*, species that commonly inhabit the respiratory tract, suggested the existence of an indigenous bacterial community. Bik *et al.* (2006) identified that the most abundant bacteria in the stomach, using the 16S rDNA clone library approach, was Proteobacteria, followed by Firmicutes, Bacteroidetes, Actinobacteria and Fusobacteria phyla. Similar findings were also reported in other studies (Delgado *et al.*, 2013; Li *et al.*, 2009; Stearns *et al.*, 2011).

There were inconsistent results between studies because of the lack of a consistent definition of normal gastric microbiota. Nevertheless, regardless of the differences in methodology (culture or 16S rRNA), sampled location (antrum or corpus biopsy specimens or fluid aspiration) or studied subjects (healthy volunteers or patients with normal endoscopic findings), the components of gastric microbiota were similar to those of the upstream digestive tract (the main composition of Firmicutes at phylum level and *Streptococcus* spp. at the genus level). However, they also had their own characteristics (the high frequency of Proteobacteria phylum) (Fig. 7.1 and Table 7.2).

Table 7.2 The composition of normal gastric microbiota

Articles	Patients	Sample type	Methods	Results
Monstein *et al.* (2000)	5 *H. pylori*-negative patients with normal gastric morphology	Temperature gradient gel electrophoresis of 16S rRNA amplicons	Corpus and antrum biopsy specimens	3 major phyla (Proteobacteria, Firmicutes and Actinobacteria)
Bik *et al.* (2006)	11 *H. pylori*-negative patients	16S rDNA clone library	Corpus and antrum biopsy specimens	5 major phyla (Proteobacteria, Firmicutes, Bacteroidetes, Actinobacteria and Fusobacteria) and 128 phylotypes
Zilberstein *et al.*(2007)	20 healthy volunteers	Culture	Aspirate	5 phyla (Firmicutes, Proteobacteria, Actinobacteria, Fusobacteria and Bacteroidetes);18 genera with the main components being *Veillonella* spp., *Lactobacillus* spp. and *Clostridium* spp.

Table 7.2 Continued

Articles	Patients	Sample type	Methods	Results
Andersson *et al*.(2008)	3 healthy volunteers	16S rRNA 454 pyrosequencing	Corpus biopsy specimens	13 phyla with 5 main components (Firmicutes, Actinobacteria, Bacteroidetes, Proteobacteria and Fusobacteria); 262 phylotypes separated into one group specific to throat (*Streptococcus*, *Actinomyces*, *Prevotella* and *Gemella* genera) and one group specific for stomach (Proteobacteria)
Li *et al*. (2009)	5 *H. pylori*-negative patients	16S rRNA sequencing	Antrum and corpus biopsy specimens	5 major phyla with the most abundant being Proteobacteria, followed by Firmicutes, Bacteroidetes, Actinobacteria and Fusobacteria
Stearns *et al*. (2011)	4 *H. pylori*-negative patients	16S rRNA Illumina sequencing	Antrum and corpus biopsy specimens	3 phyla (Proteobacteria, Firmicutes and Bacteroidetes) with antral samples; 3 phyla (Firmicutes, Proteobacteria and Fusobacteria) with corpus samples
Delgado *et al*. (2013)	4 *H. pylori*-negative patients	16S rRNA amplicons pyrosequencing and culture	Antrum biopsy specimens	Main genera *Streptococcus*, *Propionibacterium* and *Lactobacillus* by both methods

Colorectum

Colorectum has the most diverse, dense microbial community in the gut microbiota, with up to 10^{14} bacteria. The composition of this ecosystem varies widely and comprises mucosal and faecal microbiota. Despite the differences in microorganisms residing in the colorectal lumen and mucosal surface, in general the normal colorectal flora mainly consisted of Bacteroides and Firmicutes (cluster Clostridia), followed by Proteobacteria, Actinobacteria, Fusobacteria and Verrucomicrobia at phylum level (Eckburg *et al.*, 2005; Hayashi *et al.*, 2003; Lu *et al.*, 2016; Wang *et al.*, 2005). Besides genera from the phyla Firmicutes and Bacteroidetes, the human colon also harbours primary pathogens, e.g. species such as *Campylobacter jejuni, Salmonella enterica, Vibrio cholerae, Escherichia coli* and *Bacteroides fragilis,* but with a low abundance ($\leq$0.1% of the entire gut microbiome). The abundance of the phylum Proteobacteria is markedly low and its absence along with a high abundance of signature genera such as *Bacteroides, Prevotella* and *Ruminococcus* suggests a healthy gut microbiota. Besides this longitudinal difference, there also exists an axial difference from the lumen to the mucosal surface of the intestine. Whereas *Bacteroides, Bifidobacterium, Streptococcus,* Enterobacteriacae, *Enterococcus, Clostridium, Lactobacillus* and *Ruminococcus* are the predominant luminal microbial genera (can be identified in stool), only *Clostridium, Lactobacillus, Enterococcus* and *Akkermansia* are the predominant mucosa and mucus-associated genera (detected in the mucus layer and epithelial crypts of the small intestine). The complex microbiota develops in the first 2 years after delivery and keeps stable throughout the lifetime, but varies among individuals (Lozupone *et al.*, 2012). Therefore, despite the differences from confounding factors between studies, studies conducted on healthy volunteers

Table 7.3 The composition of normal colorectal microbiota

Articles	Patients	Sample type	Methods	Results
Eckburg *et al.* (2005)	3 healthy adults	16S rDNA clone library	Stool and biopsy specimens	Most of the members were Firmicutes and Bacteroidetes phyla; Proteobacteria, Actinobacteria, Fusobacteria, and Verrucomicrobia phyla were detected at low levels
Wang *et al.* (2005)	1 healthy 54-year-old female	16S rDNA clone library	Biopsy specimens	6 phyla (Firmicutes, Bacteroidetes, Proteobacteria, Fusobacteria, Verrucomicrobia and Actinobacteria); Firmicutes and Bacteroidetes were the predominant phyla (88%)
Hayashi *et al.* (2005)	6 healthy elderly individuals	16S rDNA clone library	Stool	Majority of bacteria were *Bacteroides* spp. and relatives, *Clostridium* rRNA cluster IV and IX, *Clostridium* rRNA subcluster XIVa and Gammaproteobacteria
Zilberstein *et al.*(2007)	24 healthy volunteers	Culture	Aspirate	Predominance with the presence of 'non-pathogenic' anaerobic bacteria *Veillonella* spp.
Lu *et al.* (2016)	20 healthy volunteers	16S rRNA pyrosequencing	Biopsy specimens	Firmicutes was the most predominant phylum (88.6%), followed by Bacteroidetes (5.8%) and Proteobacteria (3.3%)

showed that normal colorectal microbiota was characterized by the predominance of Furmicutes and Bacteroides and the low abundance of Proteobacteria and Fusobacteria phyla (Fig. 7.1 and Table 7.3).

Gut microbiota composition and common GI tract carcinogenesis

The composition of gut microbiota in oesophageal cancer

Oesophageal cancer is the eighth most common cancer, responsible for about 5% of cancer-related deaths worldwide (Ferlay *et al.*, 2015). Currently, it has become a significant concern because of its increasing rate during the past four decades, especially in Western countries (Pohl and Welch, 2005). The underlying aetiopathogenesis of oesophageal diseases including gastroesophageal reflux disease, Barrett's oesophagus and EC relies on the chronic, prolonged exposure of the distal oesophagus to harmful contents such as acid from the stomach or bile salts from the duodenum. These factors were thought to perturb the normal oesophageal ecosystem and trigger the progression to different disease states. However, the mechanism about the relationship between dysbiosis and the diseases has not yet been fully understood.

To date, the only study about microbiota in oesophageal adenocarcinoma using cultivation-based methods reported that there was increasing tendency in the richness or diversity of microbial compositions (56 species belonging to 19 genera in control subjects vs. 73 species belonging to 23 genera in the cancer group) (Blackett *et al.*, 2013). Yang *et al.* (2009) found that oesophageal microbiota could be divided into two types, termed type I and type II. Type I was dominated by Gram-positive bacteria (Firmicutes phylum, *Streptococcus* genus) and was closely associated with a histologically normal oesophagus. In contrast, type II was dominated by Gram-negative bacteria (Bacteroidetes, Proteobacteria, Fusobacteria and Spirochaetes phyla) and was associated with oesophagitis and Barrett's oesophagus that were precancerous lesions and thus predisposed to EC (Yang *et al.*, 2009). In addition, meta-analyses showed that *H. pylori* infection is related to the decreased risk of EC (Islami and Kamangar, 2008; Nie *et al.*, 2014).

Even though the evidence about the association between microbiota and EC are negligible, the above data emphasized the role of dysbiosis in the oesophageal tumorigenesis. To date, there was no single particular bacterial species responsible for EC. The current data suggested that the abundance of bacterial diversity with the dominance of Gram-negative anaerobic/microaerophilic bacteria might be the characteristics of microbiota in EC. Further studies with large-scale, well-designed approaches are necessary to characterize and clarify the relationship of gut microbiota and EC.

The composition of gut microbiota in gastric cancer

Gastric cancer is the fifth most common cancer and was the third most common cause of cancer-related death worldwide (Ferlay *et al.*, 2015). Since it was discovered in the early 1980s, *H. pylori* has become the most important risk factor for the development of GC. This organism is a paradigm of microbe-driven gastric diseases. Epidemiology studies showed that 74.7–89% of cases of GC, especially non-cardia GC, were infected with *H. pylori* (Plummer *et al.*, 2015; Song and Zhou, 2015). Historically, studies were focusing on *H. pylori*, believed to be the only pathogen that could be isolated from the harsh gastric environment. However, the mechanism of the remaining proportion of GC (10%) is still unknown. With the development of high-throughput sequencing technology, it allowed to focus on the role of gastric microbiota in general as well as the non-*H. pylori* members in particular to fulfil the remaining puzzle in the picture of gastric carcinogenesis (Table 7.4).

Table 7.4 The composition of gut microbiota in gastric cancer

Articles	Patients	Sample type	Methods	Results
Dicksved *et al.* (2009)	10 non-cardiac GC patients; 5 dyspeptic control patients	Antrum and corpus biopsy specimens	Terminal restriction fragment length polymorphism (T-RFLP); 16S rRNA gene cloning and sequencing	Relatively low abundance of *H. pylori* in cancer group; the microbiota of non-*H. pylori* GC cases was instead dominated by the genera *Streptococcus*, *Lactobacillus*, *Veillonella* and *Prevotella*
Han *et al.* (2010)	51 GC patients	Paraffin-embedded tissues from resected GC	16S rRNA pyrosequencing	92.2% of GC cases were *H. pylori* positive; *Helicobacter cinaedi*, *Helicobacter mustelae* and *Campylobacter hyointestinalis* were responsible for the remaining non-*H. pylori* cases

Table 7.4 Continued

Articles	Patients	Sample type	Methods	Results
Aviles-Jimenez *et al*. (2014)	5 non-atrophic gastritis (NAG); 5 intestinal metaplasia (IM); 5 intestinal-type GC patients	Antrum and tumour biopsy specimens	PhyloChip G3 microarray	Bacterial diversity decreased from NAG to IM to GC; *Porphyromonas*, *Neisseria*, TM7 group and *Streptococcus sinesis* decreased whereas *Lactobacillus coleohominis* and *Lachnospiraceae* increased from NAG to IM to GC
Eun *et al*. (2014)	10 chronic gastritis (CG), 10 IM, 11 non-cardia GC patients	Biopsy specimens	16S rRNA 454 GSFLX Titanium	Bacterial diversity increases in GC compared with IM and CG; Epsilonproteobacteria class and Helicobacteraceae family significantly decreased whereas the Bacilli class and Streptococcaceae family increased in GC, especially in the *H. pylori* dominant group
Jo *et al*. (2016)	34 GC patients; 29 controls	Antrum and corpus biopsy specimens	16S rRNA 454 pyrosequencing	No significant difference inmicrobial composition between cancer and control groups; non-*H. pylori* nitrate-reducing bacteria tend to increase in the cancer group and may have a weak role compared with *H. pylori* in gastric carcinogenesis
Wang *et al*. (2016)	6 CG, 6 GC patients	Antrum biopsy specimens	16S rRNA pyrosequencing	Bacterial richness in GC higher than in CG; *Lactobacillus*, *Escherichia/Shigella*, *Nitrospirae*, *Burkholderia fungorum* and *Lachnospiraceae* genera enriched in GC
Li *et al*. (2017)	8 *H. pylori* negative controls, 9 *H.pylori*-positive gastritis, 18 IM, 14 GC patients	Biopsy specimens	16S rRNA Illumina MiSeq	Bacterial diversity significantly decreased in normal control group compared with cancer group
Coker *et al*. (2017)	21 superficial gastritis (SG), 23 atrophic gastritis, 17 IM, 20 GC patients	Biopsy specimens	16S rRNA Illumina MiSeq	GC and IM significantly reduced the richness compared with SG; phylum level (Fusobacteria) and genus level (*Peptostreptococcus*, *Dialister*, *Mogibacterium*) were significantly enriched in GC; *Peptostreptococcus stomatis Streptococcus anginosus*, *Parvimonas micra*, *Slackia exigua*, *Dialister pneumosintes* spp.

The first study by Dicksved *et al.* (2009), in attempting to characterize the gastric microbiota in GC patients compared with dyspeptic patients, did not find any significant differences between GC and control groups. With a low abundance of *H. pylori* in cancer samples, the study mentioned about the potential role of non-*H. pylori* bacteria (*Streptococcus, Lactobacillus, Veillonella* and *Prevotella* genera) accounting for 83.3% (five out of six) of non-*H. pylori* GCs. However, the sample size in this study was small and there seemed to be selection bias because of the low prevalence of *H. pylori* infection in the sampled area. Han *et al.* (2010) showed that *Helicobacter cinaedi, Helicobacter mustelae* and *Campylobacter hyointestinalis* might be responsible for non-*H. pylori* GC cases. A recent study by Jo *et al.* (2016), focusing on nitrate-reducing bacteria, found that non-*H. pylori* nitrate-reducing bacteria tended to dominate the GC group and therefore might have a role in gastric carcinogenesis. In the non-*H. pylori* GC group, there was more bacterial diversity due to overgrowth of other bacteria. In fact, anything causing a reduction of acid secretion, such as autoimmune gastritis, a complication of acid-suppressive therapy, could create favourable conditions for the overgrowth of bacteria other than *H. pylori*.

Keeping with the global consensus by using conventional methods, the studies based on 16S rRNA also showed that *H. pylori* was presented in 49.5–92.2% of GC cases (Han *et al.*, 2010). The pathogenesis of *H. pylori*-driven cancer was deciphered by so-called 'Correa's cascade' through the progressive steps from chronic gastritis, atrophy, intestinal metaplasia, dysplasia and, eventually, gastric adenocarcinoma (Correa, 1992). Several studies confirmed the decrease of bacterial diversity towards GC, especially in the *H. pylori*-positive group during this progression (Aviles-Jimenez *et al.*, 2014; Coker *et al.*, 2018; Li *et al.*, 2017). The decline of bacterial diversity might be explained by the presence of *H. pylori* in the gastric environment that supports to shape the microbial community in this organ (Li *et al.*, 2017; Wang *et al.*, 2016). Indeed, several studies revealed that *H. pylori* could inhibit the other commensal bacteria in the gastric microenvironment (Andersson *et al.*, 2008; Llorca *et al.*, 2017; Maldonado-Contreras *et al.*, 2011). Nevertheless, the other studies revealed that the bacterial diversity increased during different stages of GC development (Eun *et al.*, 2014; Wang *et al.*, 2016). The fact that *H. pylori* was difficult to colonize in atrophic, intestinal metaplasia and GC tissues suggested that *H. pylori* infection might be more important in the early stages of gastric carcinogenesis as a trigger for the development of chronic atrophic gastritis rather than as a direct inducer of GC. Instead, other bacteria in gastric microbiota might be responsible for later stages of GC development due to the decreased acidity secondary to mucosal atrophy (Dias-Jacome *et al.*, 2016). It is similar to the 'driver-passenger' in CRC described in the next section. The opposite results between studies might be caused by sampling biases (the proportion of *H. pylori* positive in the sample or the definition of *H. pylori* positive based on operational taxonomic units), sample size or different platforms of sequencing technology (Eun *et al.*, 2014; Wang *et al.*, 2016). Regarding the role of bacterial composition in *H. pylori*-related GC, Eun *et al.* (2014) found that the Proteobacteria phylum (Epsilonproteobacteria class) significantly decreased whereas the Firmicutes phylum increased. Coker *et al.* (2018) also showed that Fusobacteria and Firmicutes phyla (*Peptostreptococcus, Dialister, Mogibacterium* genera) were significantly enriched in the GC group. Interestingly, the study also considered the potential roles of *Peptostreptococcus stomatis, Streptococcus anginosus, Parvimonas micra, Slackia exigua* and *Dialister pneumosintes* species as biomarkers to distinguish GC from different stages, even after validating these candidates into the other cohort population.

The results from microbiome studies were variable and controversial. This might be caused not only by internal factors such as the anatomical and physiological differences between gastric sites, the state of gastric acid, reflux of bile acids and gastric peristalsis, but also by external factors such as diet, medications and methodologies. To date, there was no particular bacterial signature specifically responsible for GC. Nonetheless, the role of *H. pylori* in the development of GC was evident. Current studies also indicated the role of oral bacteria (Firmicutes phylum) other than *H. pylori* contributing to the development of *H. pylori*-related GC (Fig. 7.1). The emerging evidence of the presence of non-*H. pylori* bacteria in the development of non-*H. pylori* GC is yet unclear. The tendency of bacterial diversity during the progression of GC is also still inconclusive; however, the role of dysbiosis in the pathogenesis of GC has been established and documented.

The composition of gut microbiota in colorectal cancer

Colorectal still remains a significant health issue and is the third most commonly diagnosed malignancy and the fourth leading cause of cancer death in the world, especially in high human development index countries (Arnold *et al.*, 2017). There were estimated to be 1.3 million new cases and 690,000 deaths in 2012 (GLOBOCAN 2012; http://globocan.iarc.fr). While *H. pylori* has been directly linked to GC, the role of bacteria in the colon to cause CRC remains controversial. However, bacterial load in the colon is 1 million-fold higher than in the small intestine and approximately 12-fold more cancers develop in the former than in the latter, suggesting a potential role of gut microbiota in colorectal carcinogenesis (Proctor, 2011). Indeed, with 99% of the microbial mass in the intestinal tract, the colorectal microbiota have the great impact on human health and are the most studied microbiota. Hence, it is critical to reveal the role of dysbiosis in CRC and its composition and diversity need to be characterized.

One of the earliest studies to correlate gut bacteria and CRC was conducted by Swindsinski *et al.* (1998), who reported an association between *E. coli* and CRC. O'Keefe *et al.* (2007) observed that the abundance of hydrogen sulphide and bile salt-producing bacteria was associated with increased risk of CRC. Because the majority of bacteria in the colon are not culturable, these early studies evaluated only bacteria that were culturable and, thus, underestimated the bacterial diversity. The advancements of high-throughput sequencing methods have revolutionized to give a comprehensive description of bacterial diversity and colorectal microbiota composition at different levels. The accumulative evidence showed the decline of the bacterial diversity in CRC in comparison with non-CRC subjects (Ahn *et al.*, 2013; Chen *et al.*, 2012; Gao *et al.*, 2015). The details about the bacterial composition at different levels are shown in Table 7.5. In general, the consistent results from different studies revealed that the dysbiosis caused by the abundance of Fusobacteria (*Fusobacterium nucleatum* spp.) (Ahn *et al.*, 2013; Castellarin *et al.*, 2012; Flemer *et al.*, 2017; Gao *et al.*, 2015; Kostic *et al.*, 2012; Nakatsu *et al.*, 2015; Sinha *et al.*, 2016; Warren *et al.*, 2013) and/or Proteobacteria (*Escherichia, Porphyromonas, Streptococcus, Enterococcus* genera) (Nakatsu *et al.*, 2015; Sinha *et al.*, 2016; Wang *et al.*, 2012; Warren *et al.*, 2013) and/or Bacteroidetes (*Bacteroides fragilis* spp.) and/or Firmicutes (*Clostridia* sp.) (Flemer *et al.*, 2017; Gao *et al.*, 2015; Nakatsu *et al.*, 2015; Sinha *et al.*, 2016; Wang *et al.*, 2012) phyla was associated with CRC when compared with the non-cancer healthy control group (Fig. 7.1). The most consistent findings between studies revealed that *Fusobacterium nucleatum* species might be a potential biomarker for the development, even though it did not present in all CRC

Table 7.5 The composition of gut microbiota in colorectal cancer

Articles	Patients	Sample type	Methods	Results
Sobhani *et al.* (2011)	60 CRC, 119 normal patients	Stool	qPCR	*Bacteroides* and *Prevotella* genus significantly more abundant in CRC group than in normal group
Castellarin *et al.* (2012)	11 CRC samples, 11 normal samples	Biopsy specimens	Illumina RNA-seq, qPCR	Marked abundance of *Fusobacterium nucleatum* in CRC compared with control group; confirm the same results by validating in other population with 88 matched case–control by qPCR
Kostic *et al.* (2012)	9 CRC samples, 9 normal samples	Biopsy specimens	16S rDNA 454 FLXTitanium, qPCR	*Fusobacterium nucleatum* was significantly enriched in CRC; confirm the same results by validating in other population with 95 matched case–control by qPCRand 16S rDNA pyrosequencing; enriched in Fusobacteria phylum whereas depleted in Bacteroidetes and Firmicutes phylum in CRC
Wang *et al.* (2012)	46 CRC patients, 56 healthy volunteers	Stool	16S rRNA 454 pyrosequencing	Proteobacteria phylum and genera (*Porphyromonas*, *Escherichia/Shigella*, *Enterococcus*, *Streptococcus*, *Peptostreptococcus*) significantly increased in CRC
Chen *et al.* (2012)	46 CRC patients (21 stools, 32 swabs), 56 healthy volunteers (22 stools, 34 swabs)	Biopsy specimens and rectal swab samples	16S rDNA pyrosequencing	High diversity in non-cancerous normal tissue compared with cancerous tissue; low diversity in CRC compared with control; from tissue, enriched in Lactobacillales whereas reduction in *Faecalibacterium* in CRC; from swab, reduction in *Bifidobacterium*, *Faecalibacterium* and *Blautia* whereas enriched in *Fusobacterium*, *Porphyromonas Peptostreptococcus and Mogibacterium in CRC*
Ahn *et al.* (2013)	47 CRC patients, 94 control patients	Stool	16S rDNA 454 FLXTitanium pyrosequencing	Decrease in bacterial diversity in CRC compared with control; significant increase in *Fusobacterium* genus and Bacteroidetes phylum (*Porphyromonas* genus in CRC; depletion of Firmicutes phylum (*Clostridia* class, *Coprococcus* and Lachnospiraceae family) in CRC

Table 7.5 Continued

Articles	Patients	Sample type	Methods	Results
Warren *et al*. (2013)	65 CRC patients (65 tumour and 65 control specimens)	Biopsy specimens	Illumina RNA-seq	*Fusobacterium nucleatum*, *Campylobacter showae*, *Leptotrichia hofstadii* and *Selenomonas sputigena* species over-represented in CRC
Gao *et al*. (2015)	CRC patients (31 cancerous, 20 adjacent non-cancerous, 15 proximal colon cancer, 16 distal colon cancer tissues), healthy volunteers (15 proximal colon, 15 distal colon)	Biopsy specimens	16S rRNA 454 GS FLXpyrosequencing	Low diversity in CRC compared with control; increase in Firmicutes and Fusobacteria and decrease in Proteobacteria in CRC; increase in *Lactococcus* and *Fusobacterium* whereas reduction in *Pseudomonas* and *Escherichia/Shigella* in cancerous tissues compared with adjacent non-cancerous tissues
Nakatsu *et al*. (2015)	61 control, 47 polyp and 52 CRC patients	Biopsy specimens	16S rRNA 454 GS FLXTitanium pyrosequencing	Members of oral microbiome (*Fusobacterium*, *Bacteroides fragilis*, *Gemella*, *Peptostreptococcus* and *Parvimonas*) enriched in CRC
Sinha *et al*. (2016)	48 CRC patients, 102 control patients	Stool	16S rRNA 454 RocheFLX Titanium	*Fusobacterium*, *Porphyromonas*, Clostridia and Lachnospiraceae were significantly associated with CRC
Flemer *et al*. (2017)	59 CRC, 21 polyp and 56 control patients	Biopsy specimens and stools	16S rRNA Illumina Miseq	Members of the oral microbiome (*Fusobacterium*, *Porphyromonas*, *Peptostreptococcus*, *Parvimonas*, *Bacteroides*, *Roseburia*, *Ruminococcus* and *Oscillibacter*) enriched in CRC; stratify CRC into four groups, of which groups 1 and 2 consisted of CRC with Pathogen cluster (*Fusobacterium*, *Parvimonas* and *Peptostreptococcus*) and Prevotella cluster respectively; group 3 consisted of mostly controls with Firmicutes cluster 1; and group 4 consisted of 60% CRC cases with Bacteroidetes clusters 1 and 2. Firmicutes cluster 2 was abundant in CRC in groups 3 and 4; results confirmed in two previous cohort populations

qPCR, quantitative polymerase chain reaction.

cases. However, there was variability between these studies; this might be dependent on the analysis techniques, sample type (stool vs. specimen tissue), sample location (proximal or distal colon), sample size and the controls, which were not often comparable. Therefore, it is not possible to characterize a consistent, specific cancer-associated gut microbiome. Interestingly, the latest study showed that CRC can be stratified into four co-abundance groups based on the co-occurrence networks (Flemer *et al.*, 2017), of which groups 1 and 2 consisted of only CRC individuals with a high abundance of Pathogen and Prevotella cluster, respectively; group 3 included mostly healthy controls with a high abundance of Firmicutes cluster 1; and group 4 contained mostly CRC cases with an elevated abundance of Bacteroidetes cluster 1 and Bacteroidetes cluster 2. Firmicutes cluster 2 was abundant in CRC subjects of group 3 and 4. The results were also validated in the other two previous cohort populations. These data suggested that gut microbiota might be exploited as a screening tool for people with the highest risk of developing CRC. However, the results need to be validated in other populations and longitudinal studies are required to assess the predictive value of the findings.

The data above led to the formation of two models, the keystone pathogens or 'alpha bug' and 'driver-passenger' hypotheses. In the 'alpha bug' theory, the bacteria with virulent properties cooperate with each other to create a more hostile tumour environment whereas, in the latter model, initial bacteria (drivers) damage DNA of the colorectal epithelium resulting in overgrowth of opportunistic bacteria (passengers) and, eventually, the 'drivers' are replaced gradually by 'passengers', with advantages for tumorigenesis in the new niche (Tjalsma *et al.*, 2012).

Taken together, current evidence suggested a contributory role for microbiota in CRC development. Even though there were several candidates found from many metagenomics studies, such as *Fusobacterium nucleatum*, *Bacteroides fragilis*, *Enterococcus faecalis*, *Streptococcus bovis*, *Clostridium septicum* and *E. coli*, the specific role of these individual microorganisms needs to be clarified. Rather than a single carcinogenic bacterial strain, it seems more likely that an altered community of gut microbes is associated with CRC development. Because the gut bacteria have a modifiable nature, further studies may have potential implications for managing gut bacteria for CRC treatment.

Therapeutic approach targeting the microbiota

In the previous literature, the use of stool from healthy people to treat human diseases such as food poisoning, febrile disease and typhoid fever was forst recorded in a Chinese medical book around 300–400 years AD (Zhang *et al.*, 2012). Later, Eiseman *et al.* (1958) introduced the role of faecal enemas in treating patients with refractory ulcerative colitis. Recently, faecal microbiota transplantation was approved as a standard therapy in the treatment guideline of recurrent *Clostridium difficile* infection (Surawicz *et al.*, 2013). The easiest seen from the clinical practice about the application of microbiota in treating many diseases is the use of probiotics. The accumulating current evidence has suggested the promising potentials about the application of microbiota-dependent therapies.

Despite the potential role of microbiota in treating other diseases such as refractory ulcerative colitis, Crohn's disease and irritable bowel disease, its role in preventing or treating cancers remains scarce even at the preclinical level (Zhang *et al.*, 2018). In comparison with other parts of the GI tract, the colorectum is the most studied and applied organ for the microbiota-based therapies. A study showed that probiotics such as VSL#3 affected microbial compositions but failed to reduce colitis-related CRC (Arthur *et al.*, 2013). Current studies in both clinical and animal models showed that the modulation of gut microbiota might improve the efficacy of immunotherapy (anti-PD-1, anti-PD-L1, anti-CTLA-4), and chemotherapy (Gopalakrishnan *et al.*, 2018; Matson *et al.*, 2018; Routy *et al.*, 2018). In addition to medications, dietary factors appeared to become an important factor involved in the evolution of CRC with the most reasonable and cost-effective approach by modulating the gut microbiota (O'Keefe, 2016). Berberine, a plant-derived dietary component, was showed to be able to rescue tumour formation induced by *Fn*-induced CRC (Yu *et al.*, 2015). An animal model in rats showed that the manipulation of the gut microbiota using resistant starch was a protective factor against colitis-induced CRC (Hu *et al.*, 2016). Even the initial studies gave optimistic outcomes; however, the application in practice should be kept in caution. On the other hand, a study conducted in mice showed that *Fn* promotes the chemotherapy resistance to CRC (Yu *et al.*, 2017). It suggested that the microbiota compositions should be applied cautiously in clinical practice with accounting for the balance between the improvement of their efficacy and the reducing of their adverse effects.

In the stomach, *H. pylori* eradication decreases the incidence of GC and prevents the development of GC. However, current therapy regimens for *H. pylori* have led to a high failure rate due to the rapid increase of antibiotic resistance and alternative therapies are required. Recent meta-analyses showed that probiotic supplementation in *H. pylori* therapies could alleviate the side effects and the disease-related clinical symptoms and increase the rate of successful eradication, hence reducing the risk of gastric carcinogenesis (Lü *et al.*, 2016; Zhu *et al.*, 2014).

Hence, the current data showed proof that the dietary components and products of the gut microbiota affected cancer progression and treatment response. Even the initial results proposed that microbiota-targeted therapies may be a promising strategy in prevention and treatment; however, therapeutic interferences are still in their infancy.

Limitations and future trends of gut microbiome studies

Although many available data support the hypothesis that an imbalance in the gut microbiota might be the missing piece of the puzzle of the pathogenesis of various diseases, the associative nature of the human studies makes it difficult to elucidate the real pathogenic potential of gut microbes in the development or progression of the diseases. Indeed, the inconsistency or the heterogeneity in conducting the study such as methods (culture-dependent or culture-independent approaches), population studied (children or adults, inclusion or exclusion criteria), sampled areas, sample types (biopsy specimens or faeces) become confounding factors and make it difficult to compare between microbiome studies. In other words, these factors explained why it is difficult to find a consistent result or a specific microbial signature between these studies. Well-powered studies with more homogeneous approaches using strict inclusion criteria, adequate sampling protocol and powerful molecular technologies will assist in shedding light on this issue.

In addition, although the metagenomics approach provided better phylogenetic resolution of microbial communities, this method did not take account of the viability and function of microbes. Current studies are mostly cross-sectional studies and therefore the association between gut microbiota and specific disease can be confirmed; however, the causal relationship and the dynamics between them are missing. Therefore, longitudinal studies of large cohorts are necessary and will provide insight into the role of gut microbiota composition and the progress or the onset of specific diseases. The identification of the causal or disease-promoting bacteria will pave the way to the new diagnostic method and therapeutic strategies (microbiota-modulating strategies) to complement classical treatment to prevent in the early stage or to give better prognosis.

Although it was believed that a bacterium might not be responsible for the development of GI tract cancer, there has been special interest in identifying given bacteria considered as 'oncomicrobes' that could directly contribute or cause the development of the specific diseases. It was restricted by the challenge of cultivating the anaerobic gut bacteria. Along with the advancement of culturomic analysis, it has assisted isolating and culturing previous unculturable bacteria (Lagier *et al.*, 2016). In addition, our mechanistic understanding of the interaction between such bacterial community and human health is limited because of the obvious ethical and practical issues. It required *in vitro* novel models that resemble the physiological environment of human gut such as 'organ-on-a-chip' or human–microbial crosstalk (HuMix) platform to enhance our knowledge of host–microbe interactions (Shah *et al.*, 2016). These approaches will probably fill the gaps in our understanding about the role of microbiomes in the maintenance of health and in the pathogenesis of specific disease. Future microbiome studies are indicated to identify subjects with increased GI cancer risk, to predict clinical outcome and to develop potential treatments based on microbial profiles.

Conclusion

The role of gut microbiota in the carcinogenesis of common GI tract cancers is becoming increasingly clear. It poses a challenge for researchers to apply this knowledge in order to develop new therapeutic strategies that target the microbial ecosystem on the basis of a patient's microbiota fingerprint. Consequently, in the near future the insights about the pathogenesis, diagnosis, prevention and treatment of GI tract cancers could be modified, taking into account the intimate relationship between a human being and their symbiotic bacteria. Future studies might help to explore the applicability of the microbiome in diagnostics, prognostics and potentially prevention and treatment.

Acknowledgements

This work was supported by grants-in-aid from MEXT/JSPS [grant numbers 16H05191 (YY), 18K16182 (TM) and Core-to-Core Program (YY)]. VPT and BG are PhD students supported by the MEXT Scholarship Programme.

Conflicts of interests

No potential conflicts of interest were reported by the authors.

Author contributions

VPT, BG and TM wrote the manuscript. VPT, TM and YY designed and revised the manuscript.

References

Agans, R., Rigsbee, L., Kenche, H., Michail, S., Khamis, H.J., and Paliy, O. (2011). Distal gut microbiota of adolescent children is different from that of adults. FEMS Microbiol. Ecol. 77, 404–412. https://doi.org/10.1111/j.1574-6941.2011.01120.x.

Ahn, J., Sinha, R., Pei, Z., Dominianni, C., Wu, J., Shi, J., Goedert, J.J., Hayes, R.B., and Yang, L. (2013). Human gut microbiome and risk for colorectal cancer. J. Natl. Cancer Inst. *105*, 1907–1911. https://doi.org/10.1093/jnci/djt300.

Andersson, A.F., Lindberg, M., Jakobsson, H., Bäckhed, F., Nyrén, P., and Engstrand, L. (2008). Comparative analysis of human gut microbiota by barcoded pyrosequencing. PLOS ONE *3*, e2836. https://doi.org/10.1371/journal.pone.0002836.

Arnold, M., Sierra, M.S., Laversanne, M., Soerjomataram, I., Jemal, A., and Bray, F. (2017). Global patterns and trends in colorectal cancer incidence and mortality. Gut *66*, 683–691. https://doi.org/10.1136/gutjnl-2015-310912.

Arthur, J.C., Gharaibeh, R.Z., Uronis, J.M., Perez-Chanona, E., Sha, W., Tomkovich, S., Mühlbauer, M., Fodor, A.A., and Jobin, C. (2013). VSL#3 probiotic modifies mucosal microbial composition but does not reduce colitis-associated colorectal cancer. Sci. Rep. *3*, 2868. https://doi.org/10.1038/srep02868.

Aviles-Jimenez, F., Vazquez-Jimenez, F., Medrano-Guzman, R., Mantilla, A., and Torres, J. (2014). Stomach microbiota composition varies between patients with non-atrophic gastritis and patients with intestinal type of gastric cancer. Sci. Rep. *4*, 4202. https://doi.org/10.1038/srep04202.

Biagi, E., Nylund, L., Candela, M., Ostan, R., Bucci, L., Pini, E., Nikkïla, J., Monti, D., Satokari, R., Franceschi, C., *et al.* (2010). Through ageing, and beyond: Gut microbiota and inflammatory status in seniors and centenarians. PLOS ONE *5*, e10667. https://doi.org/10.1371/journal.pone.0010667.

Bik, E.M., Eckburg, P.B., Gill, S.R., Nelson, K.E., Purdom, E.A., Francois, F., Perez-Perez, G., Blaser, M.J., and Relman, D.A. (2006). Molecular analysis of the bacterial microbiota in the human stomach. Proc. Natl. Acad. Sci. U.S.A. *103*, 732–737.

Blackett, K.L., Siddhi, S.S., Cleary, S., Steed, H., Miller, M.H., Macfarlane, S., Macfarlane, G.T., and Dillon, J.F. (2013). Oesophageal bacterial biofilm changes in gastro-oesophageal reflux disease, Barrett's and oesophageal carcinoma: Association or causality? Aliment. Pharmacol. Ther. *37*, 1084–1092. https://doi.org/10.1111/apt.12317.

Castellarin, M., Warren, R.L., Freeman, J.D., Dreolini, L., Krzywinski, M., Strauss, J., Barnes, R., Watson, P., Allen-Vercoe, E., Moore, R.A., *et al.* (2012). Fusobacterium nucleatum infection is prevalent in human colorectal carcinoma. Genome Res. *22*, 299–306. https://doi.org/10.1101/gr.126516.111.

Chen, W., Liu, F., Ling, Z., Tong, X., and Xiang, C. (2012). Human intestinal lumen and mucosa-associated microbiota in patients with colorectal cancer. PLOS ONE *7*, e39743. https://doi.org/10.1371/journal.pone.0039743.

Claesson, M.J., Cusack, S., O'Sullivan, O., Greene-Diniz, R., de Weerd, H., Flannery, E., Marchesi, J.R., Falush, D., Dinan, T., Fitzgerald, G., *et al.* (2011). Composition, variability, and temporal stability of the intestinal microbiota of the elderly. Proc. Natl. Acad. Sci. U.S.A. *108* (Suppl. 1), 4586–4591. https://doi.org/10.1073/pnas.1000097107.

Coker, O.O., Dai, Z., Nie, Y., Zhao, G., Cao, L., Nakatsu, G., Wu, W.K., Wong, S.H., Chen, Z., Sung, J.J.Y., *et al.* (2018). Mucosal microbiome dysbiosis in gastric carcinogenesis. Gut *67*, 1024–1032. https://doi.org/10.1136/gutjnl-2017-314281.

Correa, P. (1992). Human gastric carcinogenesis: A multistep and multifactorial process – First American Cancer Society Award Lecture on Cancer Epidemiology and Prevention. Cancer Res. *52*, 6735–6740.

David, L.A., Maurice, C.F., Carmody, R.N., Gootenberg, D.B., Button, J.E., Wolfe, B.E., Ling, A.V., Devlin, A.S., Varma, Y., Fischbach, M.A., *et al.* (2014). Diet rapidly and reproducibly alters the human gut microbiome. Nature *505*, 559–563. https://doi.org/10.1038/nature12820.

Delgado, S., Cabrera-Rubio, R., Mira, A., Suárez, A., and Mayo, B. (2013). Microbiological survey of the human gastric ecosystem using culturing and pyrosequencing methods. Microb. Ecol. *65*, 763–772. https://doi.org/10.1007/s00248-013-0192-5.

Dias-Jácome, E., Libânio, D., Borges-Canha, M., Galaghar, A., and Pimentel-Nunes, P. (2016). Gastric microbiota and carcinogenesis: The role of non-*Helicobacter pylori* bacteria – A systematic review. Rev. Esp. Enferm. Dig. *108*, 530–540. https://doi.org/10.17235/reed.2016.4261/2016.

Dicksved, J., Lindberg, M., Rosenquist, M., Enroth, H., Jansson, J.K., and Engstrand, L. (2009). Molecular characterization of the stomach microbiota in patients with gastric cancer and in controls. J. Med. Microbiol. *58*, 509–516. https://dx.doi.org/10.1099/jmm.0.007302-0.

Dominguez-Bello, M.G., Costello, E.K., Contreras, M., Magris, M., Hidalgo, G., Fierer, N., and Knight, R. (2010). Delivery mode shapes the acquisition and structure of the initial microbiota across multiple body habitats in newborns. Proc. Natl. Acad. Sci. U.S.A. *107*, 11971–11975. https://doi.org/10.1073/pnas.1002601107.

Eckburg, P.B., Bik, E.M., Bernstein, C.N., Purdom, E., Dethlefsen, L., Sargent, M., Gill, S.R., Nelson, K.E., and Relman, D.A. (2005). Diversity of the human intestinal microbial flora. Science *308*, 1635–1638.

Eiseman, B., Silen, W., Bascom, G.S., and Kauvar, A.J. (1958). Fecal enema as an adjunct in the treatment of pseudomembranous enterocolitis. Surgery *44*, 854–859.

Eun, C.S., Kim, B.K., Han, D.S., Kim, S.Y., Kim, K.M., Choi, B.Y., Song, K.S., Kim, Y.S., and Kim, J.F. (2014). Differences in gastric mucosal microbiota profiling in patients with chronic gastritis, intestinal metaplasia, and gastric cancer using pyrosequencing methods. Helicobacter *19*, 407–416. https://doi.org/10.1111/hel.12145.

Ferlay, J., Soerjomataram, I., Dikshit, R., Eser, S., Mathers, C., Rebelo, M., Parkin, D.M., Forman, D., and Bray, F. (2015). Cancer incidence and mortality worldwide: Sources, methods and major patterns in GLOBOCAN 2012. Int. J. Cancer *136*, E359–86. https://doi.org/10.1002/ijc.29210.

Fillon, S.A., Harris, J.K., Wagner, B.D., Kelly, C.J., Stevens, M.J., Moore, W., Fang, R., Schroeder, S., Masterson, J.C., Robertson, C.E., *et al.* (2012). Novel device to sample the esophageal microbiome – the esophageal string test. PLOS ONE *7*, e42938. https://doi.org/10.1371/journal.pone.0042938.

Flemer, B., Lynch, D.B., Brown, J.M., Jeffery, I.B., Ryan, F.J., Claesson, M.J., O'Riordain, M., Shanahan, F., and O'Toole, P.W. (2017). Tumour-associated and non-tumour-associated microbiota in colorectal cancer. Gut *66*, 633–643. https://doi.org/10.1136/gutjnl-2015-309595.

Gagliardi, D., Makihara, S., Corsi, P.R., Viana, A.d.e.T., Wiczer, M.V., Nakakubo, S., and Mimica, L.M. (1998). Microbial flora of the normal esophagus. Dis. Esophagus *11*, 248–250.

Gao, Z., Guo, B., Gao, R., Zhu, Q., and Qin, H. (2015). Microbiota disbiosis is associated with colorectal cancer. Front. Microbiol. *6*, 20. https://doi.org/10.3389/fmicb.2015.00020.

Gopalakrishnan, V., Spencer, C.N., Nezi, L., Reuben, A., Andrews, M.C., Karpinets, T.V., Prieto, P.A., Vicente, D., Hoffman, K., Wei, S.C., *et al.* (2018). Gut microbiome modulates response to anti-PD-1 immunotherapy in melanoma patients. Science *359*, 97–103. https://doi.org/10.1126/science.aan4236.

Han, H.S., Lee, K.Y., Lim, S.D., Kim, W.S., and Hwang, T.S. (2010). Molecular identification of *Helicobacter* DNA in human gastric adenocarcinoma tissues using *Helicobacter* species-specific 16S rRNA PCR amplification and pyrosequencing analysis. Oncol. Lett. *1*, 555–558. https://doi.org/10.3892/ol_00000098.

Harmsen, H.J., Wildeboer-Veloo, A.C., Raangs, G.C., Wagendorp, A.A., Klijn, N., Bindels, J.G., and Welling, G.W. (2000). Analysis of intestinal flora development in breast-fed and formula-fed infants by using molecular identification and detection methods. J. Pediatr. Gastroenterol. Nutr. *30*, 61–67.

Hayashi, H., Sakamoto, M., Kitahara, M., and Benno, Y. (2003). Molecular analysis of fecal microbiota in elderly individuals using 16S rDNA library and T-RFLP. Microbiol. Immunol. *47*, 557–570.

Hu, Y., Le Leu, R.K., Christophersen, C.T., Somashekar, R., Conlon, M.A., Meng, X.Q., Winter, J.M., Woodman, R.J., McKinnon, R., and Young, G.P. (2016). Manipulation of the gut microbiota using resistant starch is associated with protection against colitis-associated colorectal cancer in rats. Carcinogenesis *37*, 366–375. https://doi.org/10.1093/carcin/bgw019.

Human Microbiome Project Consortium (2012). Structure, function and diversity of the healthy human microbiome. Nature *486*, 207–214. https://doi.org/10.1038/nature11234.

Islami, F., and Kamangar, F. (2008). *Helicobacter pylori* and esophageal cancer risk: A meta-analysis. Cancer Prev. Res. *1*, 329–338. https://doi.org/10.1158/1940-6207.CAPR-08-0109.

Jakobsson, H.E., Jernberg, C., Andersson, A.F., Sjölund-Karlsson, M., Jansson, J.K., and Engstrand, L. (2010). Short-term antibiotic treatment has differing long-term impacts on the human throat and gut microbiome. PLOS ONE *5*, e9836. https://doi.org/10.1371/journal.pone.0009836.

Jo, H.J., Kim, J., Kim, N., Park, J.H., Nam, R.H., Seok, Y.J., Kim, Y.R., Kim, J.S., Kim, J.M., Kim, J.M., *et al.* (2016). Analysis of gastric microbiota by pyrosequencing: Minor role of bacteria other than *Helicobacter pylori* in the gastric carcinogenesis. Helicobacter *21*, 364–374. https://doi.org/10.1111/hel.12293.

Kostic, A.D., Gevers, D., Pedamallu, C.S., Michaud, M., Duke, F., Earl, A.M., Ojesina, A.I., Jung, J., Bass, A.J., Tabernero, J., *et al.* (2012). Genomic analysis identifies association of *Fusobacterium* with colorectal carcinoma. Genome Res. 22, 292–298. https://doi.org/10.1101/gr.126573.111.

Lancet (1981). Bacteria in the stomach. Lancet 2, 906–907.

Lagier, J.C., Khelaifia, S., Alou, M.T., Ndongo, S., Dione, N., Hugon, P., Caputo, A., Cadoret, F., Traore, S.I., Seck, E.H., *et al.* (2016). Culture of previously uncultured members of the human gut microbiota by culturomics. Nat. Microbiol. *1*, 16203. https://doi.org/10.1038/nmicrobiol.2016.203.

Lam, S.Y., Yu, J., Wong, S.H., Peppelenbosch, M.P., and Fuhler, G.M. (2017). The gastrointestinal microbiota and its role in oncogenesis. Best Pract. Res. Clin. Gastroenterol. *31*, 607–618.

Li, T.H., Qin, Y., Sham, P.C., Lau, K.S., Chu, K.M., and Leung, W.K. (2017). Alterations in gastric microbiota after *H. pylori* eradication and in different histological stages of gastric carcinogenesis. Sci. Rep. *7*, 44935. https://doi.org/10.1038/srep44935.

Li, X.X., Wong, G.L., To, K.F., Wong, V.W., Lai, L.H., Chow, D.K., Lau, J.Y., Sung, J.J., and Ding, C. (2009). Bacterial microbiota profiling in gastritis without *Helicobacter pylori* infection or non-steroidal anti-inflammatory drug use. PLOS ONE *4*, e7985. https://doi.org/10.1371/journal.pone.0007985.

Liu, N., Ando, T., Ishiguro, K., Maeda, O., Watanabe, O., Funasaka, K., Nakamura, M., Miyahara, R., Ohmiya, N., and Goto, H. (2013). Characterization of bacterial biota in the distal esophagus of Japanese patients with reflux esophagitis and Barrett's esophagus. BMC Infect. Dis. *13*, 130. https://doi.org/10.1186/1471-2334-13-130.

Llorca, L., Pérez-Pérez, G., Urruzuno, P., Martinez, M.J., Iizumi, T., Gao, Z., Sohn, J., Chung, J., Cox, L., Simón-Soro, A., *et al.* (2017). Characterization of the gastric microbiota in a pediatric population according to *Helicobacter pylori* status. Pediatr. Infect. Dis. J. *36*, 173–178. https://doi.org/10.1097/INF.0000000000001383.

Lozupone, C.A., Stombaugh, J.I., Gordon, J.I., Jansson, J.K., and Knight, R. (2012). Diversity, stability and resilience of the human gut microbiota. Nature *489*, 220–230. https://doi.org/10.1038/nature11550.

Lü, M., Yu, S., Deng, J., Yan, Q., Yang, C., Xia, G., and Zhou, X. (2016). Efficacy of probiotic supplementation therapy for *Helicobacter pylori* eradication: A meta-analysis of randomized controlled trials. PLOS ONE *11*, e0163743. https://doi.org/10.1371/journal.pone.0163743.

Lu, Y., Chen, J., Zheng, J., Hu, G., Wang, J., Huang, C., Lou, L., Wang, X., and Zeng, Y. (2016). Mucosal adherent bacterial dysbiosis in patients with colorectal adenomas. Sci. Rep. *6*, 26337. https://doi.org/10.1038/srep26337.

Mackie, R.I., Sghir, A., and Gaskins, H.R. (1999). Developmental microbial ecology of the neonatal gastrointestinal tract. Am. J. Clin. Nutr. *69*, 1035S–1045S. https://doi.org/10.1093/ajcn/69.5.1035s.

Maldonado-Contreras, A., Goldfarb, K.C., Godoy-Vitorino, F., Karaoz, U., Contreras, M., Blaser, M.J., Brodie, E.L., and Dominguez-Bello, M.G. (2011). Structure of the human gastric bacterial community in relation to *Helicobacter pylori* status. ISME J. *5*, 574–579. https://doi.org/10.1038/ismej.2010.149.

Mariat, D., Firmesse, O., Levenez, F., Guimarães, V., Sokol, H., Doré, J., Corthier, G., and Furet, J.P. (2009). The Firmicutes/Bacteroidetes ratio of the human microbiota changes with age. BMC Microbiol. *9*, 123. https://doi.org/10.1186/1471-2180-9-123.

Matson, V., Fessler, J., Bao, R., Chongsuwat, T., Zha, Y., Alegre, M.L., Luke, J.J., and Gajewski, T.F. (2018). The commensal microbiome is associated with anti-PD-1 efficacy in metastatic melanoma patients. Science *359*, 104–108. https://doi.org/10.1126/science.aao3290.

Monstein, H.J., Tiveljung, A., Kraft, C.H., Borch, K., and Jonasson, J. (2000). Profiling of bacterial flora in gastric biopsies from patients with *Helicobacter pylori*-associated gastritis and histologically normal control individuals by temperature gradient gel electrophoresis and 16S rDNA sequence analysis. J. Med. Microbiol. *49*, 817–822. https://doi.org/10.1099/0022-1317-49-9-817.

Nakatsu, G., Li, X., Zhou, H., Sheng, J., Wong, S.H., Wu, W.K., Ng, S.C., Tsoi, H., Dong, Y., Zhang, N., *et al.* (2015). Gut mucosal microbiome across stages of colorectal carcinogenesis. Nat. Commun. *6*, 8727. https://doi.org/10.1038/ncomms9727.

Nardone, G., Compare, D., and Rocco, A. (2017). A microbiota-centric view of diseases of the upper gastrointestinal tract. Lancet Gastroenterol. Hepatol. *2*, 298–312.

Nie, S., Chen, T., Yang, X., Huai, P., and Lu, M. (2014). Association of *Helicobacter pylori* infection with esophageal adenocarcinoma and squamous cell carcinoma: A meta-analysis. Dis. Esophagus *27*, 645–653. https://doi.org/10.1111/dote.12194.

Norder Grusell, E., Dahlén, G., Ruth, M., Ny, L., Quiding-Järbrink, M., Bergquist, H., and Bove, M. (2013). Bacterial flora of the human oral cavity, and the upper and lower esophagus. Dis. Esophagus *26*, 84–90. https://doi.org/10.1111/j.1442-2050.2012.01328.x.

O'Keefe, S.J. (2016). Diet, microorganisms and their metabolites, and colon cancer. Nat. Rev. Gastroenterol. Hepatol. *13*, 691–706. https://doi.org/10.1038/nrgastro.2016.165.

O'Keefe, S.J., Chung, D., Mahmoud, N., Sepulveda, A.R., Manafe, M., Arch, J., Adada, H., and van der Merwe, T. (2007). Why do African Americans get more colon cancer than Native Africans? J. Nutr. *137* (Suppl. 1), 175S–182S. https://doi.org/10.1093/jn/137.1.175S.

Pajecki, D., Zilberstein, B., dos Santos, M.A., Ubriaco, J.A., Quintanilha, A.G., Cecconello, I., and Gama-Rodrigues, J. (2002). Megaesophagus microbiota: A qualitative and quantitative analysis. J. Gastrointest. Surg. *6*, 723–729.

Pei, Z., Bini, E.J., Yang, L., Zhou, M., Francois, F., and Blaser, M.J. (2004). Bacterial biota in the human distal esophagus. Proc. Natl. Acad. Sci. U.S.A. *101*, 4250–4255. https://doi.org/10.1073/pnas.0306398101.

Pei, Z., Yang, L., Peek, R.M., Jr Levine, S.M., Pride, D.T., and Blaser, M.J. (2005). Bacterial biota in reflux esophagitis and Barrett's esophagus. World J. Gastroenterol. *11*, 7277–7283.

Plummer, M., de Martel, C., Vignat, J., Ferlay, J., Bray, F., and Franceschi, S. (2016). Global burden of cancers attributable to infections in 2012: A synthetic analysis. Lancet Glob. Health *4*, e609–16. https://doi.org/10.1016/S2214-109X(16)30143-7.

Plummer, M., Franceschi, S., Vignat, J., Forman, D., and de Martel, C. (2015). Global burden of gastric cancer attributable to *Helicobacter pylori*. Int. J. Cancer *136*, 487–490. https://doi.org/10.1002/ijc.28999.

Pohl, H., and Welch, H.G. (2005). The role of overdiagnosis and reclassification in the marked increase of esophageal adenocarcinoma incidence. J. Natl. Cancer Inst. *97*, 142–146.

Proctor, L.M. (2011). The Human Microbiome Project in 2011 and beyond. Cell Host Microbe *10*, 287–291. https://doi.org/10.1016/j.chom.2011.10.001.

Ringel-Kulka, T., Cheng, J., Ringel, Y., Salojärvi, J., Carroll, I., Palva, A., de Vos, W.M., and Satokari, R. (2013). Intestinal microbiota in healthy U.S. young children and adults – a high throughput microarray analysis. PLOS ONE *8*, e64315. https://doi.org/10.1371/journal.pone.0064315.

Routy, B., Le Chatelier, E., Derosa, L., Duong, C.P.M., Alou, M.T., Daillère, R., Fluckiger, A., Messaoudene, M., Rauber, C., Roberti, M.P., *et al.* (2018). Gut microbiome influences efficacy of PD-1-based immunotherapy against epithelial tumors. Science *359*, 91–97. https://doi.org/10.1126/science.aan3706.

Schroeder, B.O., and Bäckhed, F. (2016). Signals from the gut microbiota to distant organs in physiology and disease. Nat. Med. *22*, 1079–1089. https://doi.org/10.1038/nm.4185.

Schwabe, R.F., and Jobin, C. (2013). The microbiome and cancer. Nat. Rev. Cancer *13*, 800–812. https://doi.org/10.1038/nrc3610.

Sender, R., Fuchs, S., and Milo, R. (2016). Revised estimates for the number of human and bacteria cells in the body. PLOS Biol. *14*, e1002533. https://doi.org/10.1371/journal.pbio.1002533.

Shah, P., Fritz, J.V., Glaab, E., Desai, M.S., Greenhalgh, K., Frachet, A., Niegowska, M., Estes, M., Jäger, C., Seguin-Devaux, C., *et al.* (2016). A microfluidics-based *in vitro* model of the gastrointestinal human-microbe interface. Nat. Commun. *7*, 11535. https://doi.org/10.1038/ncomms11535.

Sinha, R., Ahn, J., Sampson, J.N., Shi, J., Yu, G., Xiong, X., Hayes, R.B., and Goedert, J.J. (2016). Fecal microbiota, fecal metabolome, and colorectal cancer interrelations. PLOS ONE *11*, e0152126. https://doi.org/10.1371/journal.pone.0152126.

Song, Z.Q., and Zhou, L.Y. (2015). *Helicobacter pylori* and gastric cancer: Clinical aspects. Chin. Med. J. *128*, 3101–3105. https://doi.org/10.4103/0366-6999.169107.

Stark, P.L., Lee, A., and Parsonage, B.D. (1982). Colonization of the large bowel by *Clostridium difficile* in healthy infants: Quantitative study. Infect. Immun. *35*, 895–899.

Stearns, J.C., Lynch, M.D., Senadheera, D.B., Tenenbaum, H.C., Goldberg, M.B., Cvitkovitch, D.G., Croitoru, K., Moreno-Hagelsieb, G., and Neufeld, J.D. (2011). Bacterial biogeography of the human digestive tract. Sci. Rep. *1*, 170. https://doi.org/10.1038/srep00170.

Surawicz, C.M., Brandt, L.J., Binion, D.G., Ananthakrishnan, A.N., Curry, S.R., Gilligan, P.H., McFarland, L.V., Mellow, M., and Zuckerbraun, B.S. (2013). Guidelines for diagnosis, treatment, and prevention of *Clostridium difficile* infections. Am. J. Gastroenterol. *108*, 478–498. https://doi.org/10.1038/ajg.2013.4.

Swidsinski, A., Khilkin, M., Kerjaschki, D., Schreiber, S., Ortner, M., Weber, J., and Lochs, H. (1998). Association between intraepithelial *Escherichia coli* and colorectal cancer. Gastroenterology *115*, 281–286.

Thursby, E., and Juge, N. (2017). Introduction to the human gut microbiota. Biochem. J. *474*, 1823–1836. https://doi.org/10.1042/BCJ20160510.

Tjalsma, H., Boleij, A., Marchesi, J.R., and Dutilh, B.E. (2012). A bacterial driver-passenger model for colorectal cancer: Beyond the usual suspects. Nat. Rev. Microbiol. *10*, 575–582. https://doi.org/10.1038/nrmicro2819.

Torre, L.A., Bray, F., Siegel, R.L., Ferlay, J., Lortet-Tieulent, J., and Jemal, A. (2015). Global cancer statistics, 2012. CA Cancer J. Clin. *65*, 87–108. https://doi.org/10.3322/caac.21262.

Turnbaugh, P.J., Ley, R.E., Hamady, M., Fraser-Liggett, C.M., Knight, R., and Gordon, J.I. (2007). The Human Microbiome Project. Nature *449*, 804–810. https://doi.org/10.1038/nature06244.

Walker, A.W., Ince, J., Duncan, S.H., Webster, L.M., Holtrop, G., Ze, X., Brown, D., Stares, M.D., Scott, P., Bergerat, A., *et al.* (2011). Dominant and diet-responsive groups of bacteria within the human colonic microbiota. ISME J. *5*, 220–230. https://doi.org/10.1038/ismej.2010.118.

Wang, L., Zhou, J., Xin, Y., Geng, C., Tian, Z., Yu, X., and Dong, Q. (2016). Bacterial overgrowth and diversification of microbiota in gastric cancer. Eur. J. Gastroenterol. Hepatol. *28*, 261–266. https://doi.org/10.1097/MEG.0000000000000542.

Wang, M., Ahrné, S., Jeppsson, B., and Molin, G. (2005). Comparison of bacterial diversity along the human intestinal tract by direct cloning and sequencing of 16S rRNA genes. FEMS Microbiol. Ecol. *54*, 219–231.

Wang, T., Cai, G., Qiu, Y., Fei, N., Zhang, M., Pang, X., Jia, W., Cai, S., and Zhao, L. (2012). Structural segregation of gut microbiota between colorectal cancer patients and healthy volunteers. ISME J. *6*, 320–329. https://doi.org/10.1038/ismej.2011.109.

Warren, R.L., Freeman, D.J., Pleasance, S., Watson, P., Moore, R.A., Cochrane, K., Allen-Vercoe, E., and Holt, R.A. (2013). Co-occurrence of anaerobic bacteria in colorectal carcinomas. Microbiome *1*, 16. https://doi.org/10.1186/2049-2618-1-16.

Yang, L., Lu, X., Nossa, C.W., Francois, F., Peek, R.M., and Pei, Z. (2009). Inflammation and intestinal metaplasia of the distal esophagus are associated with alterations in the microbiome. Gastroenterology *137*, 588–597. https://doi.org/10.1053/j.gastro.2009.04.046.

Yatsunenko, T., Rey, F.E., Manary, M.J., Trehan, I., Dominguez-Bello, M.G., Contreras, M., Magris, M., Hidalgo, G., Baldassano, R.N., Anokhin, A.P., *et al.* (2012). Human gut microbiome viewed across age and geography. Nature *486*, 222–227. https://doi.org/10.1038/nature11053.

Yoshioka, H., Iseki, K., and Fujita, K. (1983). Development and differences of intestinal flora in the neonatal period in breast-fed and bottle-fed infants. Pediatrics *72*, 317–321.

Yu, T., Guo, F., Yu, Y., Sun, T., Ma, D., Han, J., Qian, Y., Kryczek, I., Sun, D., Nagarsheth, N., *et al.* (2017). *Fusobacterium nucleatum* promotes chemoresistance to colorectal cancer by modulating autophagy. Cell *170*, 548–563.e16.

Yu, Y.N., Yu, T.C., Zhao, H.J., Sun, T.T., Chen, H.M., Chen, H.Y., An, H.F., Weng, Y.R., Yu, J., Li, M., *et al.* (2015). Berberine may rescue *Fusobacterium nucleatum*-induced colorectal tumorigenesis by modulating the tumor microenvironment. Oncotarget *6*, 32013–32026. https://doi.org/10.18632/oncotarget.5166.

Zhang, F., Cui, B., He, X., Nie, Y., Wu, K., and Fan, D. (2018). Microbiota transplantation: Concept, methodology and strategy for its modernization. Protein Cell *9*, 462–473. https://doi.org/10.1007/s13238-018-0541-8.

Zhang, F., Luo, W., Shi, Y., Fan, Z., and Ji, G. (2012). Should we standardize the 1,700-year-old fecal microbiota transplantation? Am. J. Gastroenterol. *107*, 1755. https://doi.org/10.1038/ajg.2012.251.

Zhu, R., Chen, K., Zheng, Y.Y., Zhang, H.W., Wang, J.S., Xia, Y.J., Dai, W.Q., Wang, F., Shen, M., Cheng, P., *et al.* (2014). Meta-analysis of the efficacy of probiotics in *Helicobacter pylori* eradication therapy. World J. Gastroenterol. *20*, 18013–18021. https://doi.org/10.3748/wjg.v20.i47.18013

Zilberstein, B., Quintanilha, A.G., Santos, M.A., Pajecki, D., Moura, E.G., Alves, P.R., Maluf Filho, F., de Souza, J.A., and Gama-Rodrigues, J. (2007). Digestive tract microbiota in healthy volunteers. Clinics *62*, 47–54.

Zivkovic, A.M., German, J.B., Lebrilla, C.B., and Mills, D.A. (2011). Human milk glycobiome and its impact on the infant gastrointestinal microbiota. Proc. Natl. Acad. Sci. U.S.A. *108* (Suppl. 1), 4653–4658. https://doi.org/10.1073/pnas.1000083107

Index

D

E

F

G

H

I

K

L

M

N

www.ingramcontent.com/pod-product-compliance
Ingram Content Group UK Ltd.
Pitfield, Milton Keynes, MK11 3LW, UK
UKHW022034041026
15591UKWH00012B/678

9 781910 190937